# CONQUEST OF ABUNDANCE

# PAUL FEYERABEND

# CONQUEST *of* ABUNDANCE

## A TALE OF ABSTRACTION

## VERSUS THE RICHNESS OF BEING

*Edited by Bert Terpstra*

THE UNIVERSITY OF CHICAGO PRESS

*Chicago and London*

The University of Chicago Press, Chicago 60637
The University of Chicago Press, Ltd., London
© 1999 by The University of Chicago
Paperback edition 2001
All rights reserved. Published 1999

10  09  08  07                                    4  5  6  7

ISBN 0-226-24533-0 (cloth)
ISBN: 0-226-24534-9 (paperback)

Frontispiece photo courtesy of Grazia Borrini-Feyerabend.

Library of Congress Cataloging-in-Publication Data

Feyerabend, Paul K., 1924–1994
    Conquest of abundance : a tale of abstraction versus the richness of being /
Paul Feyerabend ; edited by Bert Terpstra.
        p.  cm.
    Includes bibliographical references and index.
    ISBN: 0-226-24533-0-3 (cloth : alk. paper)
    1. Reality.  2. Abstraction.  I. Terpstra, Bert  II. Title.
    B3240.F483C66 1999
    110—dc21

                                            99-32398
                                            CIP

⊗ The paper used in this publication meets the minimum requirements
of the American National Standard for Information Sciences—Permanence
of Paper for Printed Library Materials, ANSI Z39.48-1992.

# CONTENTS

Preface   vii

Preface and Acknowledgments   ix

A Note on the Editing   xv

PART ONE   The Unfinished Manuscript

Introduction   3

1. Achilles' Passionate Conjecture   19

2. Xenophanes   41

3. Parmenides and the Logic of Being   60

Interlude: On the Ambiguity of Interpretations   83

4. Brunelleschi and the Invention of Perspective   89

PART TWO   Essays on the Manuscript's Themes

1. Realism and the Historicity of Knowledge   131

2. Has the Scientific View of the World a Special Status Compared

with Other Views?   147

3. Quantum Theory and Our View of the World    161

4. Realism    178

5. Historical Comments on Realism    197

6. What Reality?    206

7. Aristotle    217

8. Art as a Product of Nature as a Work of Art    223

9. Ethics as a Measure of Scientific Truth    242

10. Universals as Tyrants and as Mediators    252

11. Intellectuals and the Facts of Life    265

12. Concerning an Appeal for Philosophy    269

Index    275

# PREFACE

In an incautious moment I promised Grazia that I would produce one more collage, an entire book, no less, on the topic of *reality*—and now I am stuck with it.

I don't mind. Writing has become a very pleasurable activity, almost like composing a work of art. There is some overall pattern, very vague at first, but sufficiently well-defined to provide me with a starting point. Then come the details—arranging the words in sentences and paragraphs. I choose my words very carefully—they must sound right, must have the right rhythm, and their meaning must be slightly off center; nothing dulls the mind as thoroughly as a sequence of familiar notions. Then comes the story. It should be interesting and comprehensible, and it should have some unusual twists. I avoid "systematic" analyses: the elements hang together beautifully, but the argument itself is from outer space, as it were, unless it is connected with the lives and interests of individuals or special groups. Of course, it is already so connected, otherwise it would not be understood, but the connection is concealed, which means that strictly speaking, a "systematic" analysis is a fraud. So why not avoid the fraud by using stories right away?

The problem of reality, on the other hand, always had a special fascination for me. Why are so many people dissatisfied with what they can see and feel? Why do they look for surprises behind events? Why do they believe that, taken together, these surprises form an entire world, and why, most strangely, do they take it for granted that this hidden world is more solid, more trustworthy, more "real" than the world from which they started? The search for surprises is natural; after all, what looked like one thing often turns out to be another. But why assume that all phenomena deceive and that (as Democritus claimed) "truth lies hidden in the abyss"? [. . .]

The book on reality very slowly took shape. The working title is

*Conquest of Abundance.* The book is intended to show how specialists and common people reduce the abundance that surrounds and confuses them, and the consequences of their actions. It is mainly a study of the role of abstractions—mathematical and physical notions especially—and of the stability and "objectivity" they seem to carry with them. It deals with the ways in which such abstractions arise, are supported by common ways of speaking and living, and change as a result of argumentation and/or practical pressure. In the book I also try to emphasize the essential ambiguity of all concepts, images, and notions that presuppose change. Without ambiguity, no change, ever. The quantum theory, as interpreted by Niels Bohr, is a perfect example of that.

*Conquest of Abundance* should be a simple book, pleasant to read and easy to understand. One of my motives for writing *Against Method* was to free people from the tyranny of philosophical obfuscators and abstract concepts such as "truth," "reality," or "objectivity," which narrow people's vision and ways of being in the world.

PAUL FEYERABEND, extract from *Killing Time,* pp. 163–64, 179

# PREFACE AND ACKNOWLEDGMENTS

I have been in doubt about whether Paul would have liked to see this book in print. I know it was very dear to him, surely a "labor of love." He kept working on it for years, reading an immense variety of materials, weaving stories and arguments, paying attention to form and style. He very much wanted the book to be pleasant to read, more a piece of craft than an intellectual product. Before dying, he did not ask me to publish the book, nor did he work at it during his stay in the hospital (except for some notes on the third version of the introduction). I did ask him what he wanted me to do about the manuscript, and he said: "keep it."

While going through Paul's unanswered mail, after his death, I found a long and thoughtful letter written to him by Bert Terpstra. The letter dealt with the very subjects Paul was writing about in *Conquest of Abundance* and struck me as having been written with sincerity, intelligence, and care. I replied to the letter, mentioning the manuscript of the book. After a correspondence that lasted some months and after Bert read part of the original manuscript, I realized that the book was potentially useful, and pleasant to read, to some individuals. In fact, its very incompleteness and fragmentary nature added a layer of ambiguity to the text and openness to its meaning—qualities that Paul was far from despising and are treated somehow in the text. Bert offered to work on compiling the book out of the various materials left by Paul. He mentioned that he was going to enjoy doing it (a fact that would have been essential for Paul) and that he was going to treat the material "as the fragments of an ancient vase" (a fact that was essential to me).

I accepted Bert's offer and I am today immensely grateful to him. As he was otherwise fully employed, he worked on the book in his free time. I can only imagine the care and constancy that it must have taken, as he dealt with a large amount of material, sometimes not available simultaneously and indeed with the attention and selflessness of a restorer of an ancient vase. Likely, without Bert's impulse and help, *Conquest of Abundance* would have remained on the shelf of the Paul K. Feyerabend Archive at the University of Konstanz. The second person who played a most supportive role is Susan Abrams of University of Chicago Press. I had gotten to know and appreciate Susan via phone, fax, express mail, and e-mail throughout the months we worked together for the publication of Paul's autobiography, *Killing Time*. With insightful kindness and professionalism, Susan gave me the feeling that publishing *Conquest of Abundance* was possible, and that it was also the right thing to do. After that decision was taken, it became only a matter of compiling the book, for which I would like to acknowledge the assistance of Ms. Brigitte Uhlemann, Curator of the Paul K. Feyerabend Archive at the University of Konstanz, and of Professor Jacques Grinevald of the University of Lausanne. Both were of great help in identifying complete references from Paul's library. My final thanks go again to Susan Abrams and the staff of the University of Chicago Press, who took the book to completion with the patience and care necessary for a posthumous work.

You have in your hands an "unfinished product"—something inevitably far from what Paul would have liked to see in print. My hope, however, is that it may still give a special reading pleasure to some readers, that it may take them through a journey not unlike a music narrative, or a wondrous walk in the woods.

The book recounts some particular moments of the evolving Western culture, times when complex worldviews, overflowing with an abundance of possible interpretations of Being—and thus realities—gave way to a few abstract concepts and stereotypical accounts. One of the main consequences of this "conquest of abundance," this coming to power of crude and monolithic ideas, is the drab world some of us live in today, a world obedient only to scientific dicta and economic imperatives.

Paul does not argue for any favored interpretation of "reality," nor does he aim at convincing anyone. Rather, the reader may feel taken, through detailed pathways, to a high vista. From there, a large and wondrous landscape opens up. Our sensory and culturally cast patterns of interpretation, which provide us with habitual and usually convenient ways of understanding and living with "reality," are perceived as the filters they actually are. The humane ebullience of Homeric Gods, the stochastic regularity of elementary particles and the devoted craftsmanship of Renaissance artists, all elements of the landscape, appear as entry points of holograms. The elements create different realities while being part of them, and offering them a reading key.

As an observer of the landscape I felt a sense of spontaneous tolerance toward all cultures and worldviews, and an appreciation of their open, changeable, ambiguous borders and distinctions. Exploring such cultural openness and ambiguity, I believe, is an aspect of the book that empowers the reader. If "potentially every culture is all cultures," then "efforts to achieve peace need no longer respect some alleged cultural integrity that [may be] nothing but the rule of one or another tyrant."[1] And such awareness does not lead to detachment or cynicism, but rather, in unison with Paul, to passionate engagement. The second empowering aspect of the book is an ontological consideration. If worldviews interact with Being in a mutually creating fashion, we do affect and shape "reality." We can choose to live in a world that makes sense to us.

Readers familiar with Paul's previous writings may find in this book a quieter, more wondering attitude. They will also recognize, however, his bold connections, his impulsive remarks, and the usual stretching of any point of view away from comfortable positions. Paul was the first to subject himself to such a gymnastics of the mind. He constantly revisited and challenged his own previous work. (Authors often write the same book several times under different titles. It has been said that Paul wrote three different books under the same title—the three published editions of *Against Method*.)[2] In the last decade of his life, for instance, Paul was not at all pleased with *Science in a Free Society*, which he did not want to see reprinted. Even the detailed treatment of relativism found in *Farewell to Reason* is further extended and, shall I say, overcome by the writings collected in this volume.

This, in fact, is the main motivation that convinced me that Paul *would* like to see this book in print. He liked to let in some fresh air

1. See part I, chapter 1 of this volume, note 25.
2. Ian Hacking, "Paul Feyerabend, Humanist," *Common Knowledge* 3, no. 2 (fall 1994): 23–28.

often, in the living room of our home as in any sort of intellectual construction. This volume—unorthodox, "ambiguous," open, unfinished—lets in some of that fresh air. Even less than ever before, Paul is here not arguing nor striving to explain. By taking us through some questions and some stories, he just points at the abundance of Being, at the human openness and tolerance of ambiguity that allow us to savor it, and at the sensory and cultural filters that mediate our relation with it. As the poet Maulana Jajal al-Din gracefully expressed nearly eight centuries ago, those filters may be lived, studied, and enjoyed.[3] As Paul hints in this book, those filters may also be compared in terms of how they reduce the richness and complexity of Being, how they support, or demean, a life that makes sense to us.

A story is like water
that you heat for your bath.
It takes messages between the fire
and your skin. It lets them meet
and it cleans you!

Very few can sit down
in the middle of the fire itself,
like a salamander or Abraham.
We need intermediaries.

A feeling of fullness comes,
but usually it takes some bread
to bring it.

Beauty surrounds us,
but usually we need to be walking
in a garden to know it.

The body itself is a screen
to shield and partially reveal

---

3. The poem "Story Water" was composed in the thirteenth century A.D. by the Sufi Persian poet Maulana Jajal al-Din, better known as Rumi. The translation I quote comes from *The Essential Rumi,* trans. Coleman Barks with J. Moyne, A. J. Arberry, and R. Nicholson (New York: Harper San Francisco, 1995).

the light that's blazing
inside your presence.

Water, stories, the body,
all the things we do, are mediums
that hide and show what's hidden.

Study them,
and enjoy this being washed
with a secret we sometimes know,
and then not.

GRAZIA BORRINI-FEYERABEND, March 1999

# A NOTE ON THE EDITING

When Paul K. Feyerabend died, he left two incomplete versions of the manuscript of *Conquest of Abundance,* an incomplete third version of the introduction, and several envelopes of notes and writings meant for one or the other chapter of the book. The second version of the manuscript is a development and expansion of the first and reaches about halfway into the first version. I have assembled and merged all this available material into part I of this volume, attempting to remain as faithful as possible to the planned structure of the book and the newer versions of writings.

The introduction posed a dilemma. The three available versions were widely different and the third one was incomplete. In fact, Paul Feyerabend rewrote the introduction many times, as a way of keeping track and elaborating the meaning of the whole book while it was being written. Clearly, version three was not going to be the definitive one. Since version three was still fragmentary and this posthumous book is a combination of various versions, it seemed meaningful to attempt a combination of the various versions of the introduction as well. I have thus made a collage of the three versions, using all the available material and fragments arranged in sequence. The resulting text is probably longer, less straightforward, and more convoluted than the author intended. It is offered to the reader not as a synthesis of the book, but as a collection of "points of departure" for the stories and arguments developed in the following chapters and essays.

While Paul Feyerabend was working on the book, he also wrote several essays that dealt with the same or very similar subjects of *Conquest of Abundance.* The essays have been published in books and scientific journals or, as "column" pieces in the magazine *Common Knowledge.* A selection of such essays appears in part II of the present volume, not in their chronological order of publication but roughly following the sequence of related subjects in the manuscript of *Conquest of Abundance*. (Relation with the subjects treated in the manuscript was also the criterion for selection among Paul Feyerabend's published essays.)

I have compared the manuscript and essays and, when sentences and paragraphs were similar, used the best available formulations. One relevant essay ("Potentially Every Culture Is All Cultures," published in *Common Knowledge,* vol. 3, no. 2, fall 1994) is not included in part II, as large segments of its text are almost identical to segments of chapter 1 of the manuscript, though in a different order. As the essay appeared to be a later and more detailed treatment of the same subject, I used the material from the essay in place of the material of the manuscript for part I. The essay, however, works its way to a different conclusion than the manuscript, and I have included that conclusion in a footnote (part I, chapter 1, note 25).

Another relevant essay ("Antilogike," published in Fred D'Agostino and I. C. Jarvie, *Freedom and Rationality: Essays in Honor of John Watkins* [Dordrecht: Kluwer Academic Publishers, 1989], 185–89) is also not included in part II, as it is nearly the same as part I, chapter 3, section 2, "Antilogike."

An important essay that Feyerabend wrote in his final period, "The End of Epistemology?" (published in R. S. Cohen and Larry Laudan, eds., *Physics, Philosophy, and Psychoanalysis: Essays in Honor of Adolf Grünbaum* [Dordrecht: Kluwer, 1983], 187–204), was not included in part II, as its subject matter differs too much from that of *Conquest of Abundance.*

The reader will notice some duplication of text in parts I and II of this volume, and among the essays in part II. This has been left on purpose. In this way, the essays are not excerpted and can be read as self-contained pieces. In addition, one can follow the different lines of thought that departed from similar considerations, and can compare the conclusions developed for various occasions and publications.

I have fully documented every decision and edit I made to compose this posthumous version of *Conquest of Abundance.* The report is stored together with the original manuscripts and envelopes in the Feyerabend archive at the University of Konstanz, Germany. All decisions and edits were made in consultation with Grazia Borrini-Feyerabend.

I worked on this volume because of my own, personal rather than professional, interest in the subject, but I still would like to offer some observations as a possible reading key. In *Conquest of Abundance* Paul Feyerabend recounts some stages in the development of Western culture. He focuses, in particular, on the trend toward an increased use of abstractions and stereotypes, and a consequent disregard for particular and peculiar details. I recognized the following as underlying ideas and elements of Feyerabend's story about this trend.

*Crude dichotomies are unsuited to express subtle ontologies.* The dichotomy reality/illusion is too crude to classify the range of phenomena that are important in our lives. Each person and culture experiences various degrees of reality, but the ontologies differ among persons and cultures. Similarly the dichotomies knowledge/opinion, righteous/sinful, etc., are too crude compared to human experience.

*Our perception is shaped by language and stereotypes.* The concepts and stereotypes in our minds mold our perceptions by isolating and amplifying those aspects that fit them and relegating other aspects to oblivion. Our experienced reality is shaped by our minds. Stereotypes are limited sets of standardized interpretations of natural phenomena, human traits, art forms, etc. Perception uses stereotypes to make recognition possible, i.e., to create order out of chaos.

*Ambiguity assures the potential for change.* No concept or stereotype can ever be fully nailed down. New situations arise and reveal the ambiguities in them, new interpretations become possible, new definitions are made, new phenomena are subsumed under an existing concept, and so on. It is this very ambiguity that makes possible both personal and cultural change. We speak of cultural change when stereotype shifts exhibit an overall pattern, like the trend toward abstraction in ancient Greece (the "rise of rationality").

*Abstract theory cannot possibly express ultimate reality.* Theories or models compare projections (i.e., stereotypical perceptions stripped of many peculiar aspects) to projections (i.e., streamlined inferences of consequences from the theories or models). The match between them is an artificial construction, often made to fit using ad hoc interpretations. The belief that high theory represents ultimate reality is not justified. At most, high theory is a summary of some aspects of the response of Being to one specific and artificial approach.

*Logic is a special form of storytelling.* Logic is valid when the meanings of the terms that enter deductions are stabilized. But concepts shift in meaning from person to person and from generation to generation. It is an inherent result of the preference for mathematically and logically formulated questions and theories that scientists obtained the story of a material, "frozen" universe, uninhabited by Gods. Parmenides tells this story very concisely.

*Being responds to some approaches, but not to all.* Being is a partly yielding, partly resisting entity of unknown properties. People "create" a particular reality by developing a practice of interaction with Being (actions and perceptions) and the associated language and concepts (mental operations between actions and perceptions). Not all practices of interaction are successful, but certainly more than one exist and give meaning to the lives of the people who develop them.

While I was reading this volume, the ingredients of Feyerabend's story that I just mentioned coalesced for me into a sort of "worldview." In place of a "frozen," material universe, I could perceive an open and changeable reality, and I became able to see, and thus I was liberated from, all sorts of fixed ideas about "the way things are."

For me, this book came at the right moment. While I was working on the manuscript I left a safe and well-paying job at a large multinational company, I started my own company, and radically changed professions. It was also at the end of my first round of editing that my wife, Marlien, died as a consequence of an untreatable disease. Nothing in my life is now the same as it was when I started working on the book. *Conquest of Abundance* helped me all along, reminding me of my capacity and freedom to shape my own life.

"Work is love made visible," says Kahlil Gibran in *The Prophet.* My work on this book is my expression of love for Marlien, for life, for the world, and the expression of my gratitude to Paul Feyerabend.

BERT TERPSTRA, April 1999

# PART ONE

---

# THE UNFINISHED MANUSCRIPT

# INTRODUCTION

*They were breathless with interest. He stood with his hand on his holster and watched the brown intent patient eyes: it was for these he was fighting. He would eliminate from their childhood everything which had made them miserable, all that was poor, superstitious and corrupt. They deserved nothing less than the truth—a vacant universe and a cooling world, the right to be happy in a way they choose. He was quite prepared to make a massacre for their sakes—first the Church and then the foreigner and then the politician—even his own chief would have to go. He wanted to begin the world again with them, in a desert.*

GRAHAM GREENE, *The Power and the Glory*

T
HE WORLD WE INHABIT is abundant beyond our wildest imagination. There are trees, dreams, sunrises; there are thunderstorms, shadows, rivers; there are wars, flea bites, love affairs; there are the lives of people, Gods, entire galaxies. The simplest human action varies from one person and occasion to the next—how else would we recognize our friends only from their gait, posture, voice, and divine their changing moods? Narrowly defined subjects such as thirteenth-century Parisian theology, crowd control, late medieval Umbrian art are full of pitfalls and surprises, thus proving that there is no limit to any phenomenon, however restricted. "For him," writes François Jacob of his teacher Hovelaque, "a bone as simple in appearance as the clavicle became a fantastic landscape whose mountains and valleys could be traversed ad infinitum."[1]

Only a tiny fraction of this abundance affects our minds. This is a blessing, not a drawback. A superconscious organism would not be su-

---

Among the papers of Paul Feyerabend three different versions of this introduction were found, the last one including some incomplete and unconnected paragraphs. The introduction that is given here is a collage of all this material, put together by the editor. *Ed.*

1. *The Statue Within* (New York: Basic Books, 1988), 94.

perwise, it would be paralyzed. Sherashevsky, a mnemonist whose life is described in a fascinating monograph by A. R. Luria, was only slightly more aware of nuances than the rest of us; yet he felt obstructed at every turn. "They are so changeable," he said of human faces. "A person's expression depends on his mood and on the circumstances under which you happen to meet him. People's faces are constantly changing. It's the different shades of expression that confuse me and make it hard to remember faces." To create order, Sherashevsky decided ". . . to block off everything that wasn't essential by covering it over in [his] mind with a large canvas."[2] He consciously eliminated large sections of his world.

Sherashevsky was bothered by perceptions and he rebuilt his perception universe. Most people are helped by the perception mechanism itself: most of the blocking that shapes our lives works independently of human wishes and intentions. Still many are bothered not by perceptions but by events in nature and society, and they react accordingly—they try to "block off" what disturbs them. For them the world is still too complicated and they want to simplify it further.

Indeed, there exist situations that endanger human life and that have to be dealt with. Bacteria, viruses, ferocious animals, illnesses of all sorts, adverse geological and meteorological conditions are examples. This world is not a paradise. People need food, shelter, protection from the elements, and trying to obtain them they change their surroundings. Unfortunately this reasonable urge to make nature and society more habitable often went far beyond what was needed for survival, and even for prosperity. Ecological problems caused by humans, for instance, started already in antiquity. Yet, the urge to interfere, to eliminate, to "improve" went much further than that: it entered the domain of ritual and belief. Assuming deities that punish transgressions and reward the spreading of the faith, many religious communities tended to enforce conformity. Entire cultures and populations were eradicated in an attempt to create a uniform world not because of some adaptive disadvantage, or because they were a hindrance to the plans of some conqueror, but because their beliefs did not agree with the truth of a particular religion or philosophy.[3] [. . .]

2. A. R. Luria, *The Mind of a Mnemonist* (Cambridge: Harvard University Press, 1987), 64, 66.

3. An extreme example is in Deuteronomy 20:16–18. "The law of Moses," writes Erich Voegelin on this point (*Order and History*, vol. 1, *Israel and Revelation* [Baton Rouge: Louisiana State University Press, 1956], 375 f.) "abounds with bloodthirsty fan-

THE SEARCH FOR REALITY that accompanied the growth of Western civilization played an important role in the process of simplifying the world. It is usually presented as something positive, or an enterprise that leads to the discovery of new objects, features, relations. It is said that it widens our horizon and reveals the principles behind the most common phenomena. But this search has also a strong negative component. It does not accept the phenomena as they are, it changes them, either in thought (abstraction) or by actively interfering with them (experiment). Both types of changes involve simplifications. Abstractions remove the particulars that distinguish an object from another, together with some general properties such as color and smell. Experiments further remove or try to remove the links that tie every process to its surroundings—they create an artificial and somewhat impoverished environment and explore its peculiarities. In both cases, things are being taken away or "blocked off" from the totality that surrounds us. Interestingly enough, the remains are called "real," which means they are regarded as more important than the totality itself. Moreover, this totality is now described as consisting of two parts: a hidden and partly distorted real world and a concealing and disturbing veil around it. The dichotomy occurs not only in Western philosophy and science, it also occurs in religious contexts and it may be there combined with the dichotomy between Good and Evil. [. . .]

IN GENERAL, intellectuals are opposed to murder and bloodshed. Many of them, but by no means all, reject violence, preach tolerance, and praise argument as the only acceptable means of settling controversies. Yet their actions can be as destructive as those of their more brutal contemporaries. Certain approaches toward what has been called the problem of reality are examples. Consider the following passage, which does not speak of murder but is still remarkable because of its violence.

> Cold and austere . . . [writes Jacques Monod, molecular biologist, Nobel Prize winner and political revolutionary in an often-quoted passage] proposing no

---

tasies concerning the radical extermination of the goyim in Canaan at large, and of the inhabitants of cities in particular. And the law to exterminate the goyim is . . . motivated by the abomination of their adherence to other gods than Yahweh: the wars of Israel in Deuteronomy are religious wars. The conception of war as an instrument for exterminating everybody in sight who does not believe in Yahweh is an innovation of Deuteronomy . . ."

explanation but imposing an ascetic renunciation of all spiritual fare this idea [i.e., the idea that objective empirical knowledge is the only source of truth] was not of a kind to allay anxiety but aggravated it instead. By a single stroke it claimed to sweep away the tradition of a hundred thousand years which had become one with human nature itself. It wrote an end to the ancient animist covenant between man and nature, leaving nothing in place of that precious bond but an anxious quest in a frozen universe of solitude. With nothing to recommend it but a certain Puritan arrogance, how could such an idea win acceptance? It did not; it still has not. It has however commended recognition; but that is because, solely because of its prodigious power of performance.[4]

The destruction caused by the progress of science cannot be described more clearly. But, of course, the progress of science is supposed to be something positive and eminently humane. The questions are thus to what extent this destruction helped humanity (or a privileged part of it), how much damage was done and what is the balance. It would also be interesting to know if Monod describes a real process and how the process got started.

Monod does not speak of people, he speaks of ideas. Ideas, not industrial lobbies, "imposed"; ideas, not academic pressure groups, "swept away"; ideas, not educators, "wrote an end"—and why? Because of their "prodigious power of performance." "[W]oven together by science, living from products, [modern societies] have become as dependent upon it, as an addict on his drug."[5] It was the technological and social success of science that separated purpose and matter, destroyed animism, and turned people into addicts.

But this success is a rather late product. It did not support the atomists who preached objectivism in antiquity (in the West) and it was still lacking at the time of Galileo and Descartes, who continued the trend. Yet these early scientists were not intimidated. Surrounded by comets which appeared from nowhere, lingered for some time, and then disappeared,[6] by new stars, plagues, strange geological shapes, unknown ill-

---

4. *Chance and Necessity* (New York: Vintage Books, 1972), 170, 169.

5. Ibid., 177.

6. Maestlin, Kepler's teacher, was one of the few astronomers who treated comets as permanent objects. For Tycho Brahe they were supernatural creations. Galileo raised objections against the usual methods of determining their distance (triangulation), which assumes that comets are solid objects: triangulating a rainbow gives nonsensical results.

nesses, irrational wars, biological malformations, oddities of weather and other "monsters," they asserted the universal, "inexorable and immutable" character of the basic laws of nature.[7]

Early Chinese thinkers had taken variety at face value. They had favored diversification and had collected anomalies—instead of trying to explain them away. Many valuable astronomical and historical reports are the result of their concentration on the particular. Aristotelians emphasized the local character of regularities and insisted on a classification by multiple essences and corresponding accidents. Natural is what happens always *or almost always,* said Aristotle.[8] Scientists of an equally empirical bent, Tycho Brahe among them, regarded some anomalous events of their time as divine interventions or, as we might say today, they took cosmic idiosyncrasies seriously; others, like Kepler, ascribed them to the subjective reactions of the telluric soul, while the great Newton, for empirical as well as religious reasons, saw in them the finger of God.[9] The defenders of objectivity, on the other hand, could quote neither facts, nor products, nor a "prodigious power of performance" in their favor; they had to find support elsewhere and they did find it—in theology. It is fascinating to see how many modern ideas emerged from detailed and rather sophisticated theological debates.[10] What made their debate so influential?

Monod realized that empiricism cannot explain the origin of modern science. In this he was ahead of many of his contemporaries. Value-free knowledge, he says, is the result not of evidence, but of a *choice* which *precedes* the collection of evidence and the arrival of performance.[11] It is part of an ethic that "is the only ethic compatible with [the modern world], the only one capable, once understood and accepted, of guiding its evo-

7. Biological malformations, meteorological oddities, etc. were usually classified as monsters which, in turn, were interpreted either as warnings, or as God's ways of showing his power, or as malfunctions of nature, or as lawful events whose mechanism was still unknown. Cf. D. Wilson, *Signs and Portents* (London: Routledge, 1993). The "inexorable and immutable" character of natural laws was emphasized by Galileo (letter to Castelli, December 21, 1613), Descartes, and others. Newton opposed.

8. *De partibus animalium* 633b27 ff.

9. *Opticks,* query 31.

10. This remark does not solve the problem of how counterinductive assumptions could gain acceptance, but it shows where in history solutions might be found.

11. J. Monod, *Chance and Necessity,* 176.

lution."[12] But now we have to ask ourselves if there is such a thing as "the modern world," i.e., a uniform way of thinking and living that reaches from Harvard University to Peruvian peasants and, if there is, if it is sufficiently wonderful to make us reconfirm the decisions that allegedly led to it. Besides, how did it happen that an ethic which, rigorously applied, would have removed the incipient sciences, was accepted together with ideas whose removal it demanded?

Another point is this. Monod says the "performance" forces us to pay attention to objectivism. What kind of performance does he have in mind? The modern "universe of solitude," he says, creates "anxiety," even "aggravates" it. It makes people cling to scientific products as an "addict" clings to "his drug." It performs rather badly in these important respects, at least according to Monod. There are other side effects as well, a drastic degradation of the environment among them. What is the balance? And could a partial retention of the "ancient animist covenant between man and nature" or, to use less loaded terms, could a greater harmony between human enterprises and nature have prevented such a development?

The scientific ethic of knowledge, says Monod, "does not obtrude itself upon man; *on the contrary, it is he who prescribes it to himself.*"[13] But where in the scientific enterprise of today are the agents who freely choose one form of knowledge over another, or to use Monod's terminology, who freely make the ethics of objectivism "the *axiomatic* condition of authenticity for all discourse and all action" (orig. italics)? What we find, with very few exceptions, are intellectual leaders repeating slogans which they cannot explain and which they often violate, anxious slaves following in their footsteps and institutions offering or withdrawing money in accordance with the fashions of the day. Besides, who would have thought that a mere *decision*, a committee report of sorts, can destroy worldviews, create anxiety, and yet prevail? And who were the agents that made the decision, what prompted them to take such an extraordinary step and what powers did they use to make it stick? Monod gives no answer.

Moreover, if the "modern world" was indeed the result of an ethical decision that destroyed the effects of "a hundred thousand years" of hu-

12. Ibid., 177.
13. Ibid., italics in the original.

man existence, then some of its considerable ills can perhaps be removed by revising the decision. And if decisions alone are too weak to effect a change, then we shall have to look for support within science and outside of it just as Galileo sought support within and outside the Roman Church. And if our actions should conflict with "the evidence," we need only remember that large parts of science arose in a similar way, by running over hostile facts with the promissory note that things would soon look different—or even without any promise whatsoever.

The events described by Monod form a passing episode of a much more comprehensive trend that arranges phenomena in a hierarchy reaching from solid and trustworthy "reality" via more fleeting occurrences to entirely spurious events. [. . .]

ALMOST EVERYBODY ADMITS that there are dreams, stones, sunrises, rainbows, fleas, murders, errors—and many other things. All these events are real in the sense that they occur, are noticed, and have effects. They also have different properties and consequences in different circumstances (e.g., royal dreams have led to murder and bloodshed). Some events deceive: what looks like a repulsing stranger turns out to be a mirror's image of one's precious personality. The ancient Greeks added divinities; for them the actions of Zeus, Athena, Hermes, Aphrodite were as "real" as dreams and rainbows, which means they occurred, had distinctive properties, and affected their surroundings. However, there is no grand dichotomy, with a solid, trustworthy, genuine reality on one side and deceiving appearances on the other. Examined carefully and without prejudice the phenomena (as opposed to opinions about them) do not support such a division.

The notion of reality makes excellent sense when applied with discretion and in the appropriate context. For example, it makes good sense to distinguish between dreams and waking events. Of course a dream is not nothing, it may affect our lives. But the effects of a dreamt event differ from the effects of an event that is perceived in broad daylight. Some cultures express the difference by referring events to different layers of reality.[14]

14. Dreams may be further subdivided into those that have significance and others that are mere "fancies of the head." The distinction is found among so-called primitives (sources in L. Levy-Bruhl, *Primitive Mentality* [Boston: Beacon Press, 1966], chap. 3), as well as in some modern followers of Freud.

Yet, a notion of reality that regulates the drawing of lines between different events cannot be explained in a simple definition. A rainbow seems to be a perfectly real phenomenon. It can be seen, painted, photographed. However, we cannot run into it. This suggests that it is not like a table. It is not like a cloud either, for a cloud does not change position with the movement of the observer while a rainbow does. The discovery that a rainbow is caused by light being refracted and reflected inside water droplets reintroduces clouds together with an explanation of the peculiarities of rainbows, thus returning to them at least some of the reality of clouds. Grand subdivisions such as the subdivision real/unreal are thus much too simplistic to capture the complexities of our world. There are many different types of events, and "reality" is best attributed to an event together with a type, not absolutely.

Common sense (tribal common sense, the use of common notions in modern languages, etc.), traditional types of religion, and other well-entrenched and practically effective forms of life[15] are built in precisely this manner. They contain subtly articulated ontologies including spirits, Gods, dreams, animals, battles, rainbows, pains, etc. Each entity behaves in a complex and characteristic way which, though conforming to a pattern, constantly reveals new and surprising features and thus cannot be captured in a formula; it affects and is affected by other entities and processes constituting a rich and varied universe. In such a universe the problem is not what is "real" and what is not—queries like these don't even count as genuine questions. The problem is what occurs, in what connection, who was, is, or could be misled by the event and how. [. . .]

ON THE OTHER HAND we find the most varied groups engaged in a "search for reality." Such a search makes sense only if what is real is assumed to be hidden, not manifest. Now, there is much in this world that is hidden from us; gossip about our own deeds and habits would be an example. But a search for reality assumes that even familiar events are not what they seem to be, rather that they conceal a more genuine and solid scenario. This assumption is the connecting link between movements as different as gnosticism, classical physics, and the various aphorisms pro-

---

15. For English, cf. J. L. Austin, *Sense and Sensibilia* (Oxford: Clarendon Press, 1962), 3, 62 ff. The distinction between "traditional" and "rationalized" forms of religion is due to Max Weber, *Wirtschaft und Gesellschaft* (Tübingen: Mohr, 1925).

duced by the many well-dressed, tie-wearing bourgeois gentlemen who attempted to revolutionize poetry, the arts, the theater, and politics.

Western scientists and philosophers not only made this assumption more specific, they also formulated different versions of it. The version I would like to discuss is contained in the following three statements:

1. important ingredients of the world are concealed;

2a. the concealed ingredients form a coherent universe whose elements and motions underlie some phenomena, while other phenomena are our products entirely;

2b. because of 2a, a truthful account of this universe and of reality must be coherent and uniform;

3. human beings play an ephemeral role; they are not directly linked to reality and they cannot change it.

The assumption is found in science and in philosophy as well as in various religious movements. Gnosticism, passages of the Upanishads, Democritus, and Plato, Marxism, physical realists such as Planck and Einstein—they all share the conviction that the manifest world, the world in which human beings live, act, enjoy themselves, suffer, and die has some of the properties of a dream or an illusion. "For us who are convinced physicists" writes Einstein "the distinction between past, present, and future has no other meaning than that of an illusion, though a tenacious one."[16]

The assumption suggests that we can penetrate the concealing layers, push them aside, and perhaps remove them entirely. In a way scientists, artists, and proponents of religious renewal are like archaeologists who,

---

16. *Correspondance avec Michèle Besso*, ed. P. Speziali (Paris: Hermann, 1979), 312; cf. also 292. The trouble is that Einstein was also an empiricist. But how can experiments, which are temporal processes and, therefore, "illusions," inform us about the nature of an illusion-free "real" world? Max Planck noticed the problem. "The two statements" he writes, "'There exists a real external world that is independent of us' and 'This world cannot be known directly' together form the basis of all physics. However, they are in conflict to a certain extent and thereby reveal the irrational element inherent in physics and in every other science" ("Positivismus und reale Außenwelt," in *Vorträge und Erinnerungen* [Darmstadt: Wissenschaftliche Buchhandlung, 1945], 235. Theological debates about the nature of Christ in a similar way concern the distance between a supreme reality, God, and the world of human beings. The two-nature doctrine tries to reduce this distance while adoptionism or the idea that Christ was wholly divine and did not really suffer increases it.

having removed layers of deceptive, boring, or evil events, find unexpected and unusual treasures. The entities unearthed by science seem to have an additional advantage: being related to each other in lawful ways, they can be manipulated or predicted by using these laws. There can be new combinations of them and new entities may in this way arise in the phenomenological level. But these entities are important only if the resulting world is pleasant to live in, and if the gains of manipulations more than compensate for the losses entailed by the removal of the nonscientific layers. The objection that the entities and laws that connect them are "real" and that we must adapt to them, no matter how dismal the consequences, has no weight. Leaving the nonscientific layers in place, we would have been influenced by them just as we are now influenced by their replacement. [. . .]

VARIETY DISAPPEARS when subjected to a scholarly analysis. This is not the fault of scholars. Anyone who tries to make sense of a puzzling sequence of events, her or his own actions included, is forced to introduce ideas that are not in the events themselves, but put them in perspective. Even the discovery of an immanent structure changes the scene, for the events-as-they-are and the events-known-to-have-the-structure do not affect people in the same way. There is no escape: *understanding a subject means transforming it,* lifting it out of a natural habitat and inserting it into a model or a theory or a poetic account of it. But one transformation may be better than another in the sense that it permits or even explains what for the other transformation remains an unsolvable puzzle.

Even the simple attempt to describe may throw a veil of illusion over the world. An example are the reports of eighteenth-century travelers and naturalists which dismantle and reduce what they find while retaining the language of direct observation.[17] Here the things themselves seem to point toward a deeper, more profound but also more "objective" level. [. . .]

17. The process of scientific isolation and simplification starts already on the level of apparently unsophisticated description. Some travel descriptions, writes Mary Louise Pratt in her fascinating book *Imperial Eyes* (London and New York: Routledge, 1992), seem to emerge from a disembodied collective eye surveying an "uninhabited, unpossessed, unhistorical" landscape, "unoccupied even by the travelers themselves" (51). Linnaeus's descriptions of plants in a similar way break the connection with the plants' habitat.

THE WORLDS OF CHILDREN, adolescents, grown-ups and, for that matter, the world of all living things already contain the restrictions and the structures that are needed for a meaningful existence. Indeed we might say, in a somewhat parochial manner, that a large part of the abundance that surrounds us here on Earth arose in the attempt to conquer abundance. In the present essay I shall examine the views of individuals and groups who tried to supplement this (natural or divine) process by conscious decisions of their own.

The individuals and groups I am concerned with refused to take this abundance at face value. They denied that the world was as rich, knowledge as complex, and behavior as free as the common sense, the crafts, and the religious beliefs of their time seemed to imply. Trying to articulate their denial they introduced gross dichotomies such as real/apparent, knowledge/opinion, righteous/sinful. The early Greek philosophers and scientists especially assumed that the "real world" they had introduced in this way was simple, uniform, subjected to stable principles, and the same for all. It was not without structure—but new ways were used to determine this structure and crude alternatives served to separate it from the rest. The resulting "problems of knowledge and reality"—problems that have survived until today—were not fruits of refined ways of thinking; they arose because delicate matters had been compared with crude ideas and had been found lacking in crudeness.

We can occasionally explain why crude ideas get the upper hand: special groups want to create a new tribal identity or preserve an existing identity amidst a rich and varied cultural landscape; to do so they "block off" large parts of the landscape and either cease to talk about them, or deny their "reality," or declare them to be wholly evil.[18] For the Gnostics the entire "material world" (itself a gross simplification) was evil deception. Crude ideas often led to (limited) factual and social successes (modern example: the success of quantification in astronomy and in the physical sciences); this encouraged the proponents and reinforced their ways of thinking: ontological delicacy counts little when the survival of

18. The first way characterizes the rise of monotheism after Moses (Y. Kaufmann, *The Religion of Israel* [London: George Allen and Unwin, 1961]); the third was used, in different ways, by Deuteronomy (cf. note 3, above) and by Saint Paul (1 Corinthians 10:20). The sciences contain many examples of both the first and the second way.

a tribe, a nation, or a religious group, or the reputation of a respected profession is at stake.

The rise of philosophy and science in ancient Greece is a fascinating and rather well documented instance of the second way, the denial of reality. Although its leading representatives took great pains to separate themselves from their surroundings, their ideas were connected with these surroundings in many ways. There did exist situations that showed traces of the uniformity the early philosopher-scientists ascribed to their "real world." In politics abstract groups had replaced neighborhoods (and the concrete relationships they contained) as the units of political action (Cleisthenes); in economics money and the associated abstract notions of value had replaced barter with its attention to context and detail; the relations between military leaders and their soldiers became increasingly impersonal; local Gods merged in the course of travel, and tribal and cultural idiosyncrasies were evened out by trade, politics, and other types of international exchange; important parts of life became bland and colorless, and terms tied to specifics accordingly lost in content, or in importance, or simply disappeared. The philosophers[19] generalized these features, they increased the blandness (or, to describe the situation in a way more flattering to philosophers, the "elevation into the domain of ideas")[20] that had invaded large areas of ancient civilization, and called the result a "real world."

This was a most amazing assertion. We may grant that the new ways, being adapted to new and rather abstract procedures, had considerable merit: money increased trade, international collaboration encouraged the transfer of material and intellectual discoveries, democracy brought new strata into the political process. However, the details did not therefore cease to exist, just as people don't cease to have a nose when being weighed. Yet this was exactly what some philosophers asserted: the details, they said (or implied) were not just *irrelevant for* this or that purpose, they were *unreal* (or "subjective," to use a later term)—period—and should be disregarded. Like the rulers of Orwell's *1984* they declared less

19. I shall soon identify some of the individuals I have in mind. For the time being I call them "philosophers" without wishing to imply that all philosophers who lived during the period in question held the views and used the procedures here described.

20. G. W. F. Hegel, *Vorlesungen über die Geschichte der Philosophie* (Frankfurt: Suhrkamp, 1971), 1:290.

to be more, and more to be nonexistent. This was the most brazen denial of abundance yet pronounced.

In issuing their denial the philosophers went beyond religious and social prohibitions—they used arguments. Now argument itself was not a new invention. People had argued long before logicians started thinking about the matter. Argument, like language, or art, or ritual, is universal; and like language, or art, or ritual, it has many forms. A simple gesture, a grunt even, can decide a debate to the satisfaction of some participants while others need long and colorful arias to be convinced. Yet, what the Greek philosophers and scientists invented and then used to defend their views was not just argument. It was a special and standardized way of arguing, which, some of them thought, was independent of the situation in which it occurred, and whose results had universal validity.

The belief was mistaken in three respects. First, there is not one way of arguing, there are many (I am now speaking of logic only). One important early form of argumentation, the reductio ad absurdum, was called in doubt even as late as the beginning of the twentieth century (Brouwer). Second, the uniformity of the Real World, or of Being, could be proved only if a corresponding uniformity had already entered the premises (see chapter 2, section 5, and chapter 3 for examples). Thus, the most one can say is that the arguments that tried to establish uniformity formalized a historical process; they did not initiate it. Third, the whole approach was fundamentally incoherent. For how can what is real and not manifest be discovered, or proved, by means of what is manifest and not real? How can an objective reality that is not given be explored with the help of appearances (thoughts, perceptions, memories) that are given, but are idiosyncratic and deceptive?[21] How can information that is a result of accidental historical events be about history-independent facts and laws? Or, to use a formulation that is adapted to more recent ideas: how can humans, starting from their specific (and limited) cultural (and evolutionary) niche and relying on their specific (and very limited)

---

21. One of the earliest Western thinkers to state the paradox is Democritus. Having asserted (*Die Fragmente der Vorsokratiker*, Diels-Kranz, eds. [Zurich: Weidmann, 1985], fragment 9) that "bitter and sweet are opinions, color is an opinion, in truth there are atoms and the void," he continues (according to Galen, who preserved the passage: Diels-Kranz, fragment 125), "Wretched mind, do you take your evidence from us and then throw us down? That throw is your overthrow."

talents not only move beyond both but also describe their achievements in understandable terms? The process that is occasionally called the "rise of rationalism" certainly did not evolve in a very rational way.

But it did evolve, and it had results. Western science from Euclid and Ptolemy via Galileo, Newton, Darwin, to modern molecular biology is built on the dichotomy real/apparent. The ancient denial of abundance and the inversion of values it implied (the perceived world in which we make plans, love, hate, suffer, and the arts which try to deal with this world are on a lower plane of reality than the abstract constructions of theology, philosophy, and the sciences) affected Western civilization and, through it, the entire world. Even the most determined opponents of Western ways of life admit that science offered surprising insights and found new ways of alleviating scarcity and suffering. Moving away from common sense and common experience into a world of abstract notions did have advantages. But these advantages were distorted and turned into a menace by the basic incoherence of the enterprise. [. . .]

WITH THESE COMMENTS I return to my original problem. How is it that views that reduce abundance and devalue human existence can become so powerful? What are the processes that give them strength and make them plausible? How can we deal with this plausibility? How can we explain the radical suggestions of Parmenides, Einstein, and some molecular biologists for whom human life as a whole is an illusion? How did it happen that the rich, colorful, and abundant world that affects us in so many ways was divided into two large domains, the one still containing some life while the other lacks almost all the properties and events that make our existence important?[22]

Being confronted with questions such as these, many philosophers eschew history and take refuge in "systematic" considerations. They do not ask how the dichotomy arose, they ask how it can be justified. Thus Kant writes in the preface to his *Prolegomena:* "There are scholars for whom the history of (ancient as well as modern) philosophy already constitutes philosophy. The present prolegomena are not written for them. They must wait until those who try to use reason itself as a source have finished their considerations and they can inform the world of these events after-

---

22. Note that empirical success cannot have been the reason, for it arose and became important only much later.

wards." First the thinkers think, then the historians report what they did. This puts the cart before the horse. For the question is not what is right for a particular thinker, but what made a historical figure believe in his ideas and why they were effective. Besides, compared with "reason itself," many influential thinkers thought rather badly. According to Leucippus the void is not-Being and the void exists[23]—obvious nonsense, from a modern logical point of view. Yet the nonsense was understood and gave rise to a movement that not only lasted for over two millennia but also produced "real knowledge"—at least according to those who are now using "reason itself as a source." But if that is how knowledge arises then the question of its validity must be more closely connected with the question of its acceptance than "reason itself" would allow. And here we must be prepared for all sorts of surprises. "What we are dealing with is the *connection* and the *growth* of ideas; we are not dealing with interesting curiosities," writes Ernst Mach in the preface to his *Wärmelehre*.[24] How does he know that the "growth of ideas" is unaffected by "curiosities"? We certainly should not erect boundaries before we start. To see how this works I shall in the first chapter discuss an episode that may be seen as a particular instance of the discovery and/or invention of the real/apparent dichotomy. [. . .]

IN THE ESSAYS THAT FOLLOW I shall explore the observations made here and try to make sense of it. My procedure will be historical and episodic. I shall describe selected events and developments from the history of common sense, philosophy, science, and the arts; I shall analyze their outcome and conclude each description and analysis with a few general observations.

I will start by describing two different ways in which the distinction between appearance and reality was introduced. In the first case (see chapter 1), the distinction emerged from a complicated social process and was formulated by an unwilling participant. It was a result of pressure, not of argument. In the second case (see chapters 2 and 3) the distinction was drawn by an individual who apparently defended it by purely intellectual means. I shall show that the second case contains the first, and

23. Aristotle, *Metaphysica* 985b.
24. E. Mach, *Die Principien der Wärmelehre* (Leipzig: Johann Ambrosius Barth, 1896), v–vi; his emphasis.

that speaking of "objective progress" makes no sense in either. My next example for illustrating problems of reality and cultural change comes from the arts (see chapter 4). The example has been widely discussed and a variety of conjectures, concerning questions of the most varied kind, has been connected with it. I used some of the relevant works but I have drawn my own conclusions. The last chapter summarizes the observations and transforms them into a single story.[25] The story restricts the epistemic role of theory and extends the epistemic function of the arts but without being tied to the concepts of either. Altogether my aim is to get away from the generalizing tendencies that are found not only among scholars, but, after centuries of widespread and determined attempts at "education," also among the general public, in the West and elsewhere. The story also answers the questions raised on page 16 in the paragraph "With these comments I return to my original problem. . . ."

25. The final chapter that is mentioned here was not written. *Ed.*

# CHAPTER ONE

# ACHILLES' PASSIONATE

# CONJECTURE

*1. The Debate with the Visitors*

IN BOOK 9 OF THE *ILIAD*, Aias, Odysseus, and Phoenix, acting as messengers, ask Achilles to return to the Achaeans and to aid them in their battle against Troy. Achilles had been offended by Agamemnon, the leader of the Greeks; he had withdrawn and the situation had deteriorated. Now Agamemnon offers an enormous present and the hand of his daughter in marriage (114 ff.). For the messengers this is suitable compensation; they urge Achilles to relent. Achilles whines and splutters—and refuses. In a long speech he explains the reasons for his attitude. "Equal fate," he says "befalls the negligent and the valiant fighter; equal honor goes to the worthless and the virtuous." Striving after honor no longer makes any sense.

At this the messengers "f[a]ll silent, dismayed at his word, for he had resisted in a stunning way" (430 f.)—but they soon start arguing again. Phoenix points out that the Gods, whose powers far exceed those of humans, can be reconciled by gifts and sacrifice (497 ff.); Aias adds that even the murder of a brother or of a son has its blood price (632 f.). This is how conflicts were resolved in the past and this is how Achilles should act now. Aias ascribes Achilles' resistance to his cruelty (632). Achilles remains adamant.

Returning to the camp, Odysseus reports what has happened. Again the Greeks "f[a]ll silent, for he had spoken in a stunning way" (693 f.). They explain Achilles' attitude by his anger (679) and his pride (700). Then Diomedes suggests forgetting about Achilles and fighting without him (697 ff.).

What we have here is a rather familiar clash of attitudes—contrariness

Portions of this chapter were originally published as the article "Potentially Every Culture Is All Cultures," in *Common Knowledge* 3, no. 2 (fall 1994). Reprinted by permission of Oxford University Press.

and persistent anger on one side, surprise and a plea to be reasonable on the other. The parties try to justify their attitudes. The messengers seem to be close to commonsense while Achilles sounds a little strange.

The episode is problematic in a familiar and annoying but manageable way. The episode becomes profound and paradoxical when lifted out of its natural habitat and inserted into a model or theory. One theory that has become rather popular assumes that languages, cultures, stages in the development of a profession, a tribe, or a nation are closed in the sense that certain events transcend their capacities. Languages, for example, are restrained by rules.[1] Those who violate the rules of a language do not enter new territory; they leave the domain of meaningful discourse. Even facts in these circumstances dissolve, because they are shaped by the language and subjected to its limitations. Looking at the exchange in *Iliad* 9 with such ideas in mind, some scholars have turned it into a rather sinister affair.

Thus A. Parry[2] writes that Achilles "is the one Homeric hero who does not accept the common language . . . [He] has no language with which to express his disillusionment. Yet he expresses it, and in a remarkable way. He does it by misusing the language he disposes of. He asks questions that cannot be answered and makes demands that cannot be met." According to Parry, Achilles tries to express a situation that lies outside the limits of his language. He "can in no sense, including that of language (unlike, say, Hamlet) leave the society which has become alien to him."

Parry does not summarize the episode in *Iliad* 9, he interprets it. And he does not interpret it in accordance with the poet's scenario, but by providing a framework of his own. The framework is not arbitrary. It is based on an empirical study of the Homeric text. This text has indeed certain regularities. One of the regularities Parry has in mind is that honor and the rewards of honor cannot be separated. Now it is indeed true that "the Homeric notion of honor," to use a phrase that often occurs in this connection, is a social and not a metaphysical notion. Like other epic concepts, the concept of honor refers to an aggregate containing individual

1. Details and additional references on the peculiarities of Homeric language are found in my *Against Method* [London: Verso, 1975], chap. 17—chap. 16 of the newer editions [London: Verso, 1987 and 1993] and in *Farewell to Reason* [London and New York: Verso, 1987], 65–72, 90–103, 138–39, and 252–53).

2. A. Parry, "The Language of Achilles," *Transactions and Proceedings of the American Philological Association* 87 (1956): 6 f.

and collective actions and events. Some of the events of the aggregate are: the role (of the individual possessing or lacking honor) in battle, in the assembly, during internal dissension; his place at public ceremonies; the spoils and gifts he receives when the battle is finished; and, naturally, his behavior on all these occasions. Honor is present when (most of) the elements of the aggregate are present, absent otherwise (*Il.* 12.310 ff.— Sarpedon's speech). An explanation of honor, accordingly, would use a list, not comprehensive concepts.[3] Separating the rewards of honor from honor itself, Achilles crosses the boundaries of Homeric Greek.

This is indeed an interesting observation and the regularity which it assumes may be real. However, it does not follow that the regularity is never violated, or that it is necessary, or that it constitutes meaning so that whoever violates it is bound to talk nonsense. Perhaps a look at other regularities and at the way in which they collaborate can provide us with the missing element.

*2. The Language of Homer*     THE LANGUAGE OF THE GREEK EPIC reflects the conditions of extemporaneous composition and recital. *Memory* demanded that there be ready-made descriptions of events. The *meter* (the hexameter) required a precise metrical fit, which means that the standard elements had to be split into various forms, adapted to different cases and different positions in the verse. *Economy* demanded that, given a situation and a metrical constraint (beginning, middle, or end of a line), there be one and only one way of continuing the narration. The last demand is satisfied to a surprising degree: "If you take in the five grammatic cases the singular of all the noun-epithet formulae used for Achilles, you will find that you have forty-five different formulae of which none has, in the same case, the same metrical value."[4]

Being provided for in this manner the Homeric poet "has no interest in originality of expression, or in variety. He uses or adapts inherited formulae." He does not have a "choice, do[es] not even think in terms of choice; for a given part of the line, whatever declension case was needed,

3. Lists are not restricted to Homer. They occur in Babylonian science, in early Greek science, in commonsense thinking, and even in Plato: the first answers Socrates receives to his what-is questions are lists, not definitions.

4. M. Parry, "Studies in the Epic Technique of Oral Versemaking. I. Homer and Homeric Style," *Harvard Studies in Classical Philology* 41 (1930): 73–147, 89.

and whatever the subject matter might be, the formular vocabulary supplied at once a combination of words ready-made."[5]

Using the formulae, the Homeric poet constructed *typical scenes* by "adding the part on in a string of words in apposition."[6] Events which in later times were articulated by grammatical hierarchies are lined up like beads on a necklace. Example (*Il.* 9.556 ff.): Meleagros "lay by his wedded wife, fair Cleopatra, daughter of fair-ankled Marpessa, daughter of Euenos, and of Ides, who was the strongest of men on earth at that time—and he against Lord Phoebus Apollo took up his bow for the sake of the fair-ankled maid: her then in their halls did the father and lady mother call by the name of Alkyon because" and so on, for ten more lines and two or three more major themes before a major stop.

Homeric Greek, our scholars tell us, arranges parts and events not in hierarchies, but side by side, like pebbles in a pattern. The Homeric world, accordingly, knows no large subdivisions such as the subdivision real/apparent. Its events are all "equally real" though by no means equally influential. The dream of a king can lead to war while the actions of a soldier have only a small effect. This *paratactic feature* which parallels the absence of elaborate systems of subordinate clauses in early Greek (and which is therefore not merely an artistic artifact) explains why Aphrodite is "sweetly laughing" when in fact she complains tearfully (*Il.* 5.375) and why Achilles is "swift footed" when he is seated and talking to Priam (*Il.* 24.559). Just as late geometric pottery used the same shape for a peace-

---

5. Ibid., 230, 242.

6. T. B. L. Webster, *From Mycenae to Homer* (New York: The Norton Library, 1964), 99 f. *Formulae* occurred in Mycenaean court poetry; they can be traced to the poetry of Near Eastern courts. Here "titles of gods, kings, and men must be given correctly, and in a courtly world the principle of correct expression may be extended further. Royal correspondence is highly formal, and this formality is extended beyond the messenger scenes of poetry to the formulae used for introducing speeches. Similarly, operations are reported in terms of the operation order, whether the operation order itself is given or not, and the technique is extended to other descriptions which have no such operation orders behind them. These compulsions all derive ultimately from the court of the king, and it is reasonable to suppose that the court in turn enjoyed such formality in poetry" (Webster, *From Mycenae to Homer*, 75 f.). Standardized elements of *content* (typical scenes; the king and the nobles in war and peace; furniture; description of beautiful things) arose from the common material and social conditions of the (Sumerian, Babylonian, Assyrian, Hurrian, Hethitic, Phoenician, Mycenaean) courts. Wandering poets adapted this basic framework to the changing local circumstances.

fully grazing kid and a kid about to be eaten by a lion, merely putting its head into a lion's mouth in the second case,[7] so Aphrodite complaining is Aphrodite—and that, in the formula, is the laughing Goddess—inserted into the situation of complaining in which she participates without a change of stereotype.

Additivity structures not only the narration, but also the epic *concepts*. Lacking the coherence of what modern logicians call the connotation of a term, they do not express inherent properties (of processes, states, or events), they summarize the sequences. *Il.* 22.357—"In your breast is a heart of iron" (the dying Hector to Achilles)—describes a particular action; it does not describe character. Permanence, indicated by the word *aiei*, means that a state, or an action, or an event occurs repeatedly; we have a clustering around a person or a thing, not a property in them. *Il.* 3.60 f.—"Your heart is relentless, like an axe-blade driven by a man's strength through the timber" (Paris to Hector)—means, accordingly, that having met Hector, a person, any person, feels like having been hit by an axe. Similarly "the always watery Zephyr" of *Od.* 14.458 is a wind in whose presence things habitually get wet, i.e., they get wet in any (distinct and particular) Zephyr-situation. Features which to later readers suggest pervasive trends are presented as aggregates, extended in space and time and tied together by a suitable word.

Apparently continuous processes such as motion are dissolved into static events. In *Il.* 22.298 Achilles drags Hector along in the dust "and the dust arose around him who was dragged, and his dark hair flowed loose on either side, and in the dust *lay* his once fair head"—it *lay* in the dust, which means that the process of dragging is presented as a sequence of stills which together constitute the motion.[8] Using later terms one might say that for the poet "time is composed of moments" (Aristotle, *Physica* 239b31, describing the highly technical arguments of Zeno). Warriors engaged in battle are divided into easily separable parts: the trunk of Hippolochos rolls through the battlefield like a *log* after Agamemnon has cut off his arms and his head (*Il.* 11.146—*holmos* = round stone of cylindrical shape), the body of Hector spins like a *top* (*Il.* 14.412), the

---

7. R. Hampl, *Die Gleichnisse Homers und die Bildkunst seiner Zeit* (Tübingen: Max Niemeyer Verlag, 1952).

8. Gebhard Kurz, *Darstellungsformen menschlicher Bewegung in der Ilias* (Heidelberg: Carl Winter Universitätsverlag, 1966), 50.

head of a Gorgythion drops to one side "like a *garden poppy* being heavy with fruit and the showers of spring" (*Il.* 8.307), and so on. Language consists of words "drifting down like wintry snowflakes" (*Il.* 3.222 describing Odysseus's speech); it involves the ear and the tongue—not a unifying mind, not the abstract unity of proof. "The tongue of man is a twisty thing" says Aenaeas in his speech to Achilles (*Il.* 22.248 ff.), "there are plenty of words there, of every kind." Again we have building blocks, or atoms,[9] arranged in aggregates that fill space and are extended in time.

Some aggregates (the relentless behavior of Hector, the wetness of the Zephyr) have names, others are nameless. For example, there exists no word for the complex pattern of limbs, motions, processes that is the living human body. The epic body, therefore, differs in two ways from its successors: it is an aggregate, not a whole; and it lacks a name. (This does not mean that the human body is not conceived as a unit. However, it has the unity of an aggregate, not that of a whole which transcends and modifies its parts.)

Even functions are decentralized: the knees move or relax; the arms swiftly storm from the shoulders. A collaboration of limbs that leads to a desirable result induces a special description—no standard phrase is available for it. "To be precise, Homer does not even have any words for the arms and the legs; he speaks of hands, lower arms, upper arms, feet, calves, and thighs. Nor is there a comprehensive term for the trunk."[10] All we get is a rag doll or a puppet sewn together from relatively independent parts.

9. Cf. Webster, *From Mycenae to Homer*, chap. 9, on the similarity to the later "scientific" philosophy of nature.

10. Snell, *The Discovery of the Mind* (New York: Harper and Row, 1960), chap. 1 n. 7. Cf. also the later German edition, *Die Entdeckung des Geistes* (Göttingen: Vandenhoeck und Ruprecht, 1975), and Snell's *Gesammelte Schriften* (Göttingen: Vandenhoeck und Ruprecht, 1966). Fritz Krafft, "Vergleichende Untersuchungen zu Homer und Hesiod," *Hypomnemata*, no. 6, 1963, comments on part of the debate that surrounded Snell's views in Germany. The *Lexikon des frühgriechischen Epos*, ed. Hans Joachim Mette (Göttingen: Vandenhoek und Ruprecht, 1979–) uses some of Snell's ideas.

Snell's assertion that the living body remains unnamed is supported by the facts that *soma* (which for Snell designates the dead body) belongs to *sinomai* 'to damage', 'to fall upon'; the lion is a *sintes*, it attacks animals and people in order to kill them. *Soma*, in the case of humans, is what can be damaged when left unburied (*Od.* 11.53; 24.187); this connects directly with the fear of being destroyed as a corpse: the corpse, though dead, is still related to the man; it was *autos*, the man himself (e.g., *Il.* 1.3); "the fear of being devoured as a corpse seems to have been more terrible than the fear of death . . . the ultimate threat one can make to one's enemy is not that he will die, but that his corpse will be left to the birds and dogs." M. Nussbaum, "Psyche in Heraclitus," pt. 2, *Phronesis* 17, no. 2 (1972): 157,

The puppet has no "soul": there is no concept that covers all the events and abilities now called "mental" or "spiritual" and no common nature or substratum is assigned to them. Dreams, sudden surges of energy, outbursts of anger enter the body, briefly reside in it, perhaps produce some action, and then leave. "Man is an open target of a great many forces which impinge on him, and penetrate his very core."[11] "His experiences are no property of his soul, anchored in deep solitude or in a formless beyond kindred to the soul, but a portion of the world."[12]

But although things, persons, processes lacked the kind of unity they acquired in later times, they were neither isolated nor ruled by chance. On the contrary—the aggregates of the Homeric world were connected more firmly than the harmonic wholes that eventually replaced them. Complex and well-defined relations joined nature, humans, and the Gods; rich clusters or spectra of terms articulated these relations and expressed the many ways in which people acquired knowledge: the "openness" of individuals was more than compensated by the ties between them. "In the Iliad . . . man is completely part of this world."[13]

with citations. Later on *soma*, via this fear, was extended to the live body—but the step apparently does not occur in Homer. Cf. Krafft, "Vergleichende Untersuchungen," 25 ff.

11. Snell, *Discovery*, 117.

12. W. F. Otto, *The Homeric Gods* (New York: Pantheon, 1954), 177. Cf. H. Fränkel, *Wege und Formen frühgriechischen Denkens* (Munich: C. H. Beck'sche Verlagsbuchhandlung, 1968), 168: "The person has no impenetrable membrane and a God is not at all something alien. Forces freely enter a human being." Snell in addition assumes that "in Homer we never find a personal decision, a conscious choice made by the acting human being—a human being who is faced with various possibilities never thinks: it now depends on me, it depends on what I decide to do" (*Schriften*, 18). For him a passage like *Il.* 9.410 ff. does not imply that Achilles is about to choose a particular path; it shows that he will eventually find himself on one of various possible paths and, having been given its description in advance, will know what to expect (cf. also Bruno Snell, *Die alten Griechen und wir* [Göttingen: Vandenhoeck und Ruprecht, 1962], 48, on terms expressing activity). This view has been criticized, most recently by Bernard Williams in his Sather lectures, Berkeley 1989, and it is perhaps too extreme: increased contextualism does not mean that the elements make no contribution. What remains is that "Homeric Man" has less coherence and is less separated from (or better integrated into) his surroundings than the "autonomous individual" of later times. See the text below.

13. H. Fränkel, *Early Greek Philosophy and Poetry* (Oxford: Oxford University Press, 1975), 80. According to Erich Auerbach, *Mimesis* (Bern: Francke, 1964), 8, the Homeric style "presents the phenomena well-articulated, tangible and visible and well-determined in all their spatial and temporal relations. Inner processes do not differ in this respect: none of their parts must remain concealed or undescribed. Without remainder,

"This world" gradually changed during the late geometric period.[14] People became "autonomous," their relations more distant and problematic. Linguistically, the change is rather obvious: the clusters of epistemic terms that had characterized the preceding period shrink, the surviving "words . . . become impoverished in content, they . . . become one-sided and empty formulae."[15] New disciplines, epistemology especially, tried to connect, in theory and with insufficient means, what had become separated in practice: the "Discovery of Mind," the rise of Western science and philosophy, the associated reflections on the nature of knowledge, the impoverishment of thought and language—all these processes were part of one and the same overall development. The development announces itself in Achilles' response to his visitors and underlies the later separation of appearance and reality.

Now the question is: did the development involve a radical break so that we have first sense, then nonsense, and then a gradual and somewhat miraculous emergence of a new sense, or did Achilles' language already contain escape routes and instruments for the construction of novelties? Also, is language the only medium for presenting what is and for introducing what can be, or are there other possibilities? And further, were the features of the Homeric language an aesthetic artifice without

---

well-disposed even in their strongest emotions the persons of Homer reveal their inside in speech; . . . no speech is so full of fear and anger that it would lack the instruments of linguistic or logical articulation or would become disordered . . . Everywhere the individual elements of the phenomena are related to each other in the clearest possible way . . . yielding uninterrupted and effortlessly flowing connections."

14. Cultural periods in ancient Greece, i.e., "sets of interlocking life styles" (O. Murray, *Early Greece* [London: Collins and Son, 1980], chap. 12), are named after the prevalent artistic style. From 1100 B.C. to 800 B.C. this was the geometric style, identified mainly on pottery and vase paintings (cf. Gisela Richter, *A Handbook of Greek Art* [London and New York: Phaidon, 1974], chaps. 4 and 11). Increasing contact with the Near East and the rise of city-states dissolved the fairly homogeneous geometric period.

15. Kurt von Fritz, *Philosophie und sprachlicher Ausdruck bei Demokrit, Platon und Aristoteles* (Darmstadt: Wissenschaftliche Buchgesellschaft, reprint, 1966), 11. Cf. also von Fritz, *Grundprobleme der Geschichte der antiken Wissenschaft* (Berlin and New York: de Gruyter, 1971), 78. Some of the clusters referred to above are described in Bruno Snell's pathbreaking essay *Die Ausdrücke für den Begriff des Wissens in der vorplatonischen Philosophie* (Berlin: Weidmannsche Buchhandlung, 1924); cf. also Snell's analysis of "optical" terms in *Entdeckung*, 13 ff. R. B. Onians, *The Origins of European Thought* (1951; reprint, Cambridge: Cambridge University Press, 1988), discusses the material side of mind and perception.

factual relevance, or did they reflect the nature of things as seen, felt, acted upon by the inhabitants of contemporary Greece? And, assuming the latter, were the changes that led away from Homeric objects changes in the things themselves, or were they changes of outlook unaccompanied by corresponding changes of the world? And where is the boundary—if there is a boundary—between a collective outlook and "the world"? Many cultures assume such a boundary, but they set it in different places. Divine appearances once were real—they are mere fantasies today. Where shall we, who examine the phenomenon, set the boundary? Note that I am inquiring about an old episode, not about a modern belief. Many "educated citizens" take it for granted that reality is what scientists say it is and that other opinions may be recorded, but need not be taken seriously. But science offers not one story, it offers many[16]; the stories clash and their relation to a story-independent "reality" is as problematic as the relation of the Homeric epics to an alleged "Homeric world."

Describing the Homeric way of looking at things I already used realistic terms. A Homeric warrior, I said, was an aggregate, not a whole. Motion, I said, was a succession of independent events, not a continuous process. Language, I said, lost details when humans increased in wholeness. Are such expressions didactic devices, or were Homeric warriors indeed transit stations—put together from relatively independent parts—of events such as dreams, Gods, anger, and so on?

*3. The World of Homer* INFLUENTIAL WRITERS have supported the second alternative. According to Benjamin Lee Whorf languages shape ideas, their grammar contains worldviews and linguistic change is accompanied by a change of facts. More recent authors concur. According to Michael Baxandall, "[A]ny language, not only humanist Latin, [the language Baxandall is concentrating on] is a conspiracy against experience in the sense of being a collective attempt to simplify and arrange experience into manageable parcels . . . To exercise a language regularly on some area of experience or activity, however odd one's motives may be, [therefore] overlays the field after a time with

16. See "The Power of Science," section 3 of "Realism and the Historicity of Knowledge," essay 1 in part II of the present volume.

a certain structure; the structure is that implied by the categories, the lexical and grammatical components of the language."[17]

These are correct and very pertinent observations. What has to be added is that language is not the only "conspiracy" and that a conspiracy-free "experience" that can be arranged "into manageable parcels" does not exist. Humans paint, produce films and videos; they dance, dream, and make music; they engage in political action, exchange goods, perform rituals, build houses, start wars, act in plays, try to please patrons—and so on. All these activities occur in a fairly regular way; they contain patterns, "press" the practitioners "to conform" and in this way mold their thought, their perception, their actions, and their discriminative abilities.[18]

Music, for example, shapes moods, it makes them stand out from an otherwise unstructured background and thus in a sense creates them; money moves attention away from the concrete and personal features which make gifts so valuable and which are still emphasized in barter; masks and caricatures imprint stable stereotypes on the continuum of facial expressions, they "overlay" and reshape individual facial expressions and thus contribute to creating the standard face of a particular person[19]; democracy disregards personal relations and concentrates on the abstract rights and duties of a "citizen"—and so on. The agencies that cause the changes can be, and

17. Baxandall, *Giotto and the Orators* (Oxford: Clarendon Press, 1971), 44, 47. Baxandall describes natural languages and special forms of discourse, such as the discourse arising from the humanists' attempt to revive classical Latin. For Whorf, cf. *Language, Thought, and Reality* (Cambridge: MIT Press, 1956). Chapter 17 of the first edition and chapter 16 of the third edition of my book *Against Method* contain a more detailed account of Whorf's theory.

18. "Because a degree of regularity and simplicity is necessary if we are to be understood," writes Baxandall, "and because also the language itself has been deeply involved in our acquiring ways of discriminating at all, a system of language is always pressing us to conform with it" (*Giotto and the Orators*, 44). The same is true of other systems, such as the "system" of "discriminating" between faces, i.e., of recognizing them and of properly (for a certain culture) reacting to their expressions. Not every facial expression that shocks, pleases, or terrifies has a name and the names that do exist evoke different responses entirely.

19. Artistic characterization often uses standard types: the child, the young man, the lady of the house, the mature warrior, etc. In Greece the interest in individual features started from extremes (Centaurs, Satyrs, Thersites in the *Iliad*) and slowly moved toward more subtle presentations: B. Schweizer, *Studien zur Entstehung des Porträts bei den Griechen*, vol. 2 of *Ausgewählte Schriften* (Tübingen: Wasmuth, 1963), 115 ff. Types and extremes returned during the Roman Empire, the former for ideological purposes (cf. D. Strong, *Roman Art* [Harmondsworth: Penguin Books, 1982], 80 ff., on the Augustean period, and André Grabar, *Christian Iconography* [Princeton: Princeton University Press,

often are, described in words. However, the descriptions do not function as the agencies do, they do not "conspir[e] against experience" in the manner of the things and processes described. A wink, a stern countenance, a note sung by a teacher in front of a pupil unable to place his or her voice have results that cannot be duplicated by any set of propositions.

The existence of antagonistic "conspiracies" was recognized by the defenders of religious and political views. Iconoclasts knew that images might distort the basic message of their creed (which consisted of words and resided in Holy Books). Church architecture and church music were adapted to the needs of the Holy Faith.[20] Alternative styles were either fought or made part of religious PR. I conclude that our "field of experience" is molded, "overlaid," and "conspired against" not just by language, but by numerous other patterns and institutions, many of them in mutual conflict. An inference from a style, a particular linguistic apparatus, or, more recently, from scientific beliefs, to a cosmology, corresponding ways of life and an all-embracing "spirit of the age" therefore needs special support; it cannot be made as a matter of course.[21] For example, we have to show that the structure of the Homeric epics is not a poetic artifact but recurs in other areas as well.

---

1968], esp. 96 f.), the latter to give vent to public dissatisfaction. The cycle was repeated many times. For the Renaissance, cf. John Pope Hennessy, *The Portrait in the Renaissance* (New York: Pantheon Books, 1966).

20. A striking example is the way in which Abbot Suger of Saint-Denis adapted the structure and the ornaments of his church to the ideas of Pseudo-Dionysius Areopagita. Details in E. Panofsky, *Abbot Suger on the Abbey Church of St.-Denis and Its Art Treasures* (Princeton: Princeton University Press, 1946), and Otto von Simson, *The Gothic Cathedral* (New York: Bollingen Foundation, 1956), pt. 2. The effects of the Tridentine reform on painting and architecture are described in A. Blunt, *Artistic Theory in Italy 1450–1600* (Oxford and London: Oxford University Press, 1940), chap. 8. For the effects of music cf. *The New Oxford History of Music*, vol. 4, *The Age of Humanism* (Oxford: Oxford University Press, 1968).

21. Writers quite naturally regard language as a more fundamental structuring agency than, say, dance, and within that restricted domain often put philosophy and science above poetry. The hierarchy is not the result of study; it comes from the attempt to impose a special form of life. As usual the attempt is accompanied by crude classifications (such as the classification of all processes into mental and physical and the further classification of mental processes into thought, imagination, emotion) and an arbitrary ranking of the classes obtained. "Music," says Kant, "has the lowest position among the arts . . . because it merely plays with the emotions" (*Kritik der Urteilskraft*, Ausgabe B, 221). Ancient writers were better informed and less arbitrary. According to Aristotle (*Politica* 1339a11 ff.) music, being based on imitation, "has a power of producing a certain effect on the moral character of the soul," while Aristoxenus divided musical cognition into a variable and a stable

The task is difficult, but not impossible. The agencies that shape a form of life leave their traces not only in language but also in artworks, buildings, customs, learned treatises. Thus, if the features (additivity, lack of coherent wholes, etc.) I described in the previous section can also be found in statuary and in painting; if the Gods, nature, and humans had analogous properties in popular sayings as well as in common law; if powerful ideas such as the ideas of courage, wisdom, justice, piety (which occurred not only in Homer but turned up in public speeches and were analyzed in philosophical writings, mocked in comedy, referred to on funeral stones and other inscriptions) had Homeric and not, say, Platonic characteristics; if religion was opportunistic rather than exclusive, permitting alien Gods to enter at the drop of a hat; if the Gods were not merely revered and talked about but perceived, and perceived not just by unbalanced outsiders but by the most levelheaded representatives of the culture; if different explanations of startling events were used side by side without any feeling of discomfort; if a narrator (e.g., Herodotus) assembled but did not unify, told stories but did not use a single style; if some thinkers called the resulting information *polymathi'e*, i.e., plentiful but scattered pieces of knowledge, and tried to replace those scattered pieces by a single coherent story; if people were in the habit of answering what-is questions with lists, not with definitions, and if philosophers tried to correct that habit—then we can assume that we are dealing with an influential and relatively uniform way of life and we may expect that the people involved in it temporarily lived in a world of the kind expressed in their poems, tales, sayings, and pictures.

In chapter 17 of *Against Method* (chapter 16 of the revised editions) and again in chapter 3, section 4, of *Farewell to Reason* I argued that the classical and archaic periods were preceded by, and contained, a common sense of the kind I just described and that the language of the Homeric epics on the whole agreed with it: the epics satisfy the conditions I enumerated and thus provide the outlines of a specific world (which we may *now* call, interchangeably, the Homeric world or the late geometric world). *Viewed from the inside* this world was indeed inhabited by crea-

element (cf. the excerpt from the *Harmonic Elements* as published in O. Strunk, ed., *Source Readings in Music History* [New York: W. W. Norton and Co., 1950], 27). Both show how music limits abundance and thus qualifies as a "collective attempt to simplify and arrange experience into manageable parcels." (See also Baxandall, *Giotto and the Orators*, 44, 47). Kant's dictum was rejected by Alfred Einstein, who also examined its theoretical background: *Music in the Romantic Era* (New York: Norton, 1947), 337 ff.

tures which people with Homeric perceptions but a twentieth-century vocabulary would call anthropomorphic Gods, puppetlike humans, "objective" dreams wandering from Gods to humans and back again; it was a rich pattern of events that contained individuals as parts, not as outside observers. For those who lived the pattern it was a real world—and it may well have been the only world they knew.

But now the problem alluded to by A. Parry arises with renewed force: given this world—how did people ever get out of it? How did they manage to forget or to overcome the order that constituted their lives and gave it meaning? Was the Homeric-geometric world simply destroyed so that chaos temporarily raised its head or was it gradually transformed? And, if the latter, was it transformed by arbitrary and senseless (in the sense of this world) processes such as boredom or forgetfulness or by entering existing but as yet unused paths? Was the transformation unconscious, rising to consciousness only after major steps had been taken, or was it carried out in the full awareness of the changes implied? Can we agree with Nietzsche, who wrote, in his usual bombastic style: "No fashion helped them [the philosophers—according to Nietzsche it was they who effected the transition] and paved their way. Thus they formed what Schopenhauer, in opposition to a republic of scholars called a republic of men of genius: one giant calls out to another across the desolate intervals of time and the lofty exchange between minds continues undisturbed by the noisy doings of the midgets *[Gezwerge]* that crawl beneath them."[22] Or with Plato who spoke more calmly of "the ancient battle between philosophy and poetry" (*Republic* 607b6 f.), implying an overt fight between two professions, not a gradual and perhaps subterranean development? Should we accept the claim of early philosophers such as Xenophanes, Parmenides, and Heraclitus and of their modern admirers that they single-handedly overcame the errors of tradition, just using the power of their amazing minds?

It is clear that these questions and paradoxes depend on the assumption, stated in section 1, that languages and, with them, worlds and worldviews are closed in the sense that they admit, even constitute, some actions, thoughts, perceptions, while others are not merely excluded but rendered nonexistent. Given this assumption the change of worldviews will indeed cause major upheavals.

I already said that this assumption is part of an interpretation appara-

22. "Die Philosophie im tragischen Zeitalter der Griechen," *Werke*, ed. K. Schlechta (Munich: Ullstein Verlag, 1969), 3:1063.

tus some scholars used to understand differences between cultures and difficulties experienced by an individual traveler from one culture into another. I also said that understanding is impossible without such a (not always explicit) interpretation apparatus. But the particular apparatus we are invited to apply and which contains the assumption of artificial cultural compartments is very implausible indeed. It is of relatively recent origin and the conceptual barriers it postulates did not and still do not affect the commerce between cultures. Misunderstandings can happen. Even the most ordinary events baffle some people, enrage others, and render still others speechless. But we also find that ordinary people, i.e., people not yet confused by higher learning, readily accept statements which sound strange to their neighbors and nonsensical to scholars.[23] It is true that people can spread falsehoods and be at cross-purposes with one another, that historians and sociologists can produce chimeras. But this is not because they are hindered by conceptual barriers but because they make mistakes, because they don't pay attention, misread some very plain facts or passages and are reluctant to examine the ideas that guide them in their research. Physicians, teachers, laborers, missionaries in so far unknown cultures, astronomers interested in unity, they all constantly face new situations, products, challenges, and they deal with them, often successfully. They do not excuse themselves by saying "this is beyond the semantic boundaries of the language I am speaking."[24]

I agree that if discourse is defined as a sequence of clear and distinct propositions (actions, plans, etc.) which are constructed according to pre-

23. For examples, see Carlo Ginzburg, *The Cheese and the Worms: The Cosmos of a Sixteenth-Century Miller*, trans. John and Anne Tedeschi (New York: Penguin Books, 1982; first published in Italian, *Il formaggio e i vermi: il cosmo di un mugnaio del '500* [Torino: G. Einaudi, 1976]), and Emmanuel Le Roy Ladurie, *Montaillou: The Promised Land of Error*, trans. Barbara Bray (New York: Vintage Books, 1979; first published in French, *Montaillou: Village occitan de 1294 à 1324* [Paris: Gallimard, 1975]). Already in 1552, Copernicanism was part of Florentine gossip, which found ways of diffusing arguments against it; details appear in Leonardo Olschki, *Geschichte der neusprachlichen wissenschaftlichen Literatur*, vol. 2, [1922; reprint Vaduz: Kraus Reprint, 1965], 134 ff.). Some aspects of Florentine public life during the quattrocento implied rather unusual views about personal identity. An example is Brunelleschi's joke on Manetto di Jacopo Ammanatini (analyzed by Decio Gioseffi in "Realtà e conoscenza nel Brunelleschi," *La Critica dell'Arte* 85 [March 1965]: 8 ff.). People who accept the Resurrection, the Virgin Birth of Christ, who believe in the miracle stories of the *Legenda Aurea*, and who take the Bible literally, as did many outstanding British scientists of the nineteenth century, the young Darwin included, are not likely to be stopped by "linguistic boundaries."

24. Cf. also *Against Method*, 3d ed., 17, 204 (2d ed., 17, 214).

cise and merciless rules, then discourse has a very short breath indeed. Such a discourse would be often interrupted by "irrational" events and soon be replaced by a new discourse for which its predecessor is nonsense pure and simple. If the history of thought depended on a discourse of this kind, then it would consist of an ocean of irrationality interrupted, briefly, by mutually incommensurable islands of sense. If, on the other hand, the elements of an argument, a worldview, a culture, a theoretical framework are allowed some leeway, so that they either keep their identity through very drastic changes—in which case one might say that they have potential meanings that are actualized in different ways—or change their content without violating the worldview to which they belong, then we have no reason to assume that our ways of conveying meaning have any limits. On the contrary, we can now search for features that connect the "inside" of a language with its "outside," and thus reduce conceptually induced blindness to the real causes of incomprehension, which are ordinary, normal, run-of-the-mill inertia, dogmatism, inattention, and stupidity. I am not denying differences between languages, art forms, customs. But I would ascribe them to accidents of location and/or history, not to clear, unambiguous, and immobile cultural essences: *potentially every culture is all cultures.*[25]

25. The argument I have presented in this rather abstract way is developed with passion, wit, and many examples in Renato Rosaldo's *Culture and Truth* (Boston: Beacon Press, 1993). Rosaldo is describing classical objectivist anthropology, which not only postulates closed systems but also tries to clean them up: "Most anthropological studies of death eliminate emotions by assuming the position of the most detached observer" (15). Aiming at the discovery of strict rules that guide behavior like a juggernaut, objectivist studies "make it difficult to show how social forms can be both imposed *and* used spontaneously" (58). They fail to recognize "how much of life happens in ways that one neither plans nor expects" (91). Boundary problems, not central events, teach us about the full resources of a culture. At the boundaries, writes Gloria Andaluza, a Chicana lesbian whom Rosaldo quotes, a person

> copes by developing a tolerance for contradictions, a tolerance for ambiguity. She learns to be Indian in Mexican culture, to be Mexican from an Anglo point of view. She learns to juggle cultures. She has a plural personality, she operates in a pluralistic mode—nothing is thrust out, the good, the bad and the ugly, nothing rejected, nothing abandoned. Not only does she sustain contradictions, she turns the ambivalence into something else. (216)

It is not clear that (Rosaldo speaking) "in the present postcolonial world, the notion of an authentic culture as a autonomous internally coherent universe no longer seems tenable, except perhaps as a 'useful fiction' or a revealing distortion?" (217).

The situation is no different in the sciences. Despite a persistent fog of objectivism and despite the relativistic tricks inspired by Kuhn's idea of a paradigm, many scientists have lived and are still living with ambiguity and contradiction. They could not possibly

Another objection to the assumption, stated in section 1, is that texts, collections of artworks, and cultural periods lack the uniformity needed for the monolithic view here discussed. Parry, for example, was soon criticized for his streamlining of Homer.[26] Apparently uniform periods dissolve on close inspection. During the reign of Amenophis IV, Egyptian sculptors changed very quickly from a stern formalism to a loose naturalism and back again. (Cf. *Against Method*, rev. ed., chap. 16). The problem of Antal's magnificent *Florentine Art* (Harvard University Press, 1986) is the variety of styles in fifteenth-century Florence—and so on. Examples such as these suggest that the unity that emerged from the past

---

live in any other way. New problems need new approaches. But new approaches do not fall like manna from the heaven of creativity. Old ideas continue to be used, they are slowly twisted around until some orderly minds perceive an entirely new structure, with new limits of sense, and start doing what they do best—they nail it down. This, incidentally, is the reason why the presentation of scientific *results* differs so drastically from what happens during *research*, i.e., while people are still *thinking*, and gives such a misleading picture of it. Of course, ideas can get stuck; imagination can be dimmed by dogma, financial pressures, education, and boredom. If that happens, then the idea of a closed system with precise concepts and rules slavishly followed will appear to be the only correct representation of Thought. But that situation should be avoided, not praised.

To my mind, the most important consequence of the new attitude toward cultures that underlies Rosaldo's book is that practices that seem legitimate when referred to a closed framework cease to be sacrosanct. If every culture is potentially all cultures, then cultural differences lose their ineffability and become *special and changeable manifestations of a common human nature*. Authentic murder, torture, and suppression become ordinary murder, torture, and suppression, *and should be treated as such*. Feminism has tasks not only in the United States, but even more so in Africa, India, and South America. Efforts to achieve peace need no longer respect some alleged cultural integrity that often is nothing but the rule of one or another tyrant. And there is much reason to suspect some of the ingredients of the ideology of political correctness.

But, in making use of this new freedom of action, we must be careful not to continue old habits. Objective judgments are out; so is an abstract and ideology-driven protection of cultures. Drastic interventions are not excluded but should be made *only after* an extended contact, not just with a few "leaders," but with the populations directly involved. Having discarded objectivity and cultural separation and having emphasized intercultural processes, those who perceive medical, nutritional, environmental problems or problems of human or, more specifically, female rights have to start such processes on the spot *and with due attention to the opinions of the locals*. There exist movements that already proceed in this particularizing, nonobjective manner. Liberation theology and some approaches in the area of development are examples. Let us support these movements and learn from them instead of continuing old-style epistemologies and other "authentic" games.

26. Cf. Hugh Lloyd-Jones, "Becoming Homer," *New York Review of Books*, March 5, 1992.

section may be an illusion prompted by emphasizing similarities over differences and neglecting deviations, instead of making them the starting point of an attempt to break up the apparent uniformity. It may be an illusion created by the very views it is supposed to support. And we should never forget the tensions that exist among various areas of human activity such as art, thought, politics, and other domains. Combining ambiguity with stylistic diversity and intracultural conflicts we arrive at a view that seems to lead to fewer paradoxes than the assumption that underlies Parry's interpretation. Let us see how this works in the present case!

*4. Transitions*     ACHILLES AND HIS VISITORS talk about honor.
Achilles was offended by Agamemnon, who had taken his gifts. The offense created a conflict between what Achilles received and what he thought was his due. The visitors agree that there is a conflict. Their suggestions illustrate its customary resolution; the gifts were returned untouched, more gifts were promised, honor is restored: "With gifts promised return, the Achaeans will honor you as they would an immortal," says Phoenix (602 f.: cf. 519, 526). He adds that honor is partly constituted by gifts: "if you entered the murderous battle without any gifts, then even a victory would not bring you the same honor" (604 f.). This supports its social interpretation.

Achilles is not reassured. Extending the conflict beyond its suggested resolution he perceives a lasting clash between honor and its rewards: honor and the actions that establish and/or acknowledge its presence *always* diverge.

A brief look at the rest of the epic shows indeed that Achilles' remarks do not come out of the blue.[27] They arise from a situation—the conflict between custom and Agamemnon's actions—that lies squarely within the common sense of the time. Sensitized by his anger, Achilles remembers that merit was disregarded not only in his case but in other cases as well, and he generalizes: Honor is an orphan (318 f.). The starting point of this generalization (the description of Agamemnon's actions) conforms to the archaic notion of honor; so do the cases Achilles remembers. The traditional concept allowed for discrepancies and identified

27. Cf. *Farewell to Reason*, 268–70, for a discussion of the change of language and meaning introduced by Achilles.

them by using a standard. The full generalization—honor and its re-wards *always* diverge—severs the connection between the standard and the events that gave it substance, at least in the opinion of some scholars.

Achilles goes further. He implies that the general injustice he notices lies in the nature of things. Using modern terms, we can formulate this implication by saying that the traditional standards are no longer parts of social practice. Yet they continue to play a role. This is the first indica-tion of a dichotomy that was soon to assume considerable importance—the dichotomy between (rich, concrete, but misleading) appearances and a (simple, abstract, almost empty, but still very important) reality. And this is also the reason why some scholars say that Achilles' speech does not make sense: a general rift between appearance and reality does not fit into "the Homeric worldview."

But Homeric thought was not unprepared for grand subdivisions. Di-vine knowledge and human knowledge, divine power and human power, human intention and human speech (an example mentioned by Achilles himself: 312 f.) were opposed to each other in ways that resemble the dis-tinction Achilles is using. One might say that having cut the social links of honor, Achilles strengthens the ties of honor to divine judgment, es-pecially to the judgment of Zeus (607 f.). Such ties already existed; the judgment of the Gods always played an important social role. Even the exclusive relevance of divine judgment hinted at by Achilles was pre-pared by the eminence of the Gods and the steadily increasing power of one particular divinity—Zeus—in whom "all lines converge."[28] Consid-ered in retrospect, it seems that the situation described by Achilles was there all along, though buried in a complex net that tied divine actions to human actions and human actions to each other. Achilles identifies the situation, lifts it out of its surroundings, and simplifies it by trimming

---

28. While the judgment of Zeus has a certain amount of arbitrariness and while di-vine decisions may vary from case to case, the division adumbrated by Achilles has the clarity and power of an objective and person- (God-)independent law. Even this feature has an analogy in the epic: the willfulness of the Gods is never absolute: it is restricted by *moira,* an "irrefragable order . . . which . . . exists independently of them" and by norms which they may violate but which can be used to criticize and to judge their behavior. (Walter F. Otto, *The Homeric Gods* [New York: Pantheon Books, 1954], 42, 276. Cf. also Burkert, *Griechischen Religion der archaischen und klassischen Epoche* [Stuttgart: Walter Kohlhammer, 1977], 205 f., and F. M. Cornford, *From Religion to Philosophy* [New York: Harper and Row, 1965], 16. Reference to norms occurs implicitly in the *Iliad,* e.g., in *Il.* 24:33 ff., explicitly in tragedy, for example in Sophocles, *Antigone,* 456).

some social connections. Even this last action is not arbitrary, or "creative," for Achilles has "inductive evidence" for the weakness and, perhaps, irrelevance of the connections that he trims. Nor is he left without standards, for the judgment of the Gods remains, both for him and his visitors. What we have, then, in book 9, is a change of emphasis supported by reasons and driven by Achilles' anger. We are a long way from the disaster announced by Parry and systematized by the champions of incommensurability.

Still, we may ask if the change of focus corresponded to and was perhaps supported by some more general tendency. Had Achilles or the poet who composed his lines lived in the seventh or sixth centuries B.C., I could have answered: There was a relevant tendency, closely connected with social developments. By that period, abstract groups had replaced neighborhoods (and the concrete relationships they embodied) as the units of political action (Cleisthenes); money had replaced barter with its attention to context and detail; the relations between military leaders and their soldiers had become increasingly impersonal; local Gods had merged in the course of travel, which increased their power but reduced their humanity; tribal and cultural idiosyncrasies had been evened out by trade, politics, and other types of international exchange; important parts of life had become bland and colorless, and terms tied to specifics accordingly had lost in content or in importance, or had simply disappeared.[29] (In section 2 I hinted at some of these developments.) I could have added that individual human actions (such as the actions of Solon, of Cleisthenes, of their associates) played a large role in the process, but not with these later results as their aim. Seen "from the outside," we have an adaptation of one "conspiracy"[30] ("Homeric common sense") to others (the newly emerging structures I have just described). Seen "from the inside," we have a discovery: important features of the world are being revealed. They are not revealed "as such," but only with respect to the problem that sensitizes Achilles: the existence of personal qualities detached from the efforts of an individual or the reactions of his peers.

29. For the contraction of the rich spectra of perceptual terms, see Bruno Snell, *Die Entdeckung des Geistes: Studien zur Entstehung des europäischen Denkens bei den Griechen* (Göttingen: Vandenhoeck und Ruprecht, 1975), chap. 1.

30. "[A]ny language . . . is a conspiracy against experience in the sense of being a collective attempt to simplify and arrange experience into manageable parcels." Baxandall, *Giotto and the Orators*, 44.

But Achilles or the poet who described his actions did not live in the seventh or sixth century. He spoke at a time when the developments I enumerated were in their infancy. They had started but not yet produced their more obvious results. Achilles' speech contributed to the development and thus contains an element of invention. The invented features were part of a slowly rising structure, which means that Achilles also made a discovery. Subjectivity certainly played a role; it was Achilles' anger that made him resonate to what others did not yet notice. What he saw in a sense was already there—the judgment of the Gods was always more decisive than that of mortals—which means that Achilles' vision had an "objective core." But it is still "subjective," for the move toward increasing abstractness and the related separation of reality and appearance were not the only developments.

As becomes clear from funeral inscriptions, passages of comedy, sophistic debates, medical and historical treatises, from the unwanted lists Socrates received to his what-is questions, and from Aristotle's recommendation of precisely such lists (cf. *Politica* 1260b24 ff.), the view that things, ideas, actions, processes are aggregates of (relatively independent) parts, that giving an account means enumerating instances, not subsuming them under a single term, retained its popularity right into the classical age of Greece. Geometric thought was a seed without a well-defined genetic program; accompanied by an ever-increasing cacophony of political, philosophical, military, artistic debates, it grew into many different plants. Its concepts had precisely the ambiguity I described toward the end of the last section. Nowhere in this process do we find the breaks, the lacunae, the unbridgeable chasms suggested by the idea of closed domains of discourse.

Compare this account with Parry's. According to Parry concepts have clear meanings and the languages they form are held together by more or less well-defined rules. This is what the evidence tells him. But the evidence on which he relies is limited. It is taken from stable periods of linguistic usage and it disregards what happens during periods of change. As a result certain moves become "linguistically impossible" and, if they do occur, difficult to explain.[31] Now if we add the forbidden moves to the

---

31. Explanations do exist and I have used some of them in *Farewell to Reason*, 270. It *is* possible to postulate limits of sense and a corresponding incommensurability. I myself have defended such an approach.

evidence, then the rules that seemed to forbid them lose their restrictive power. But they had this power before, when they separated sense from nonsense. I conclude that the rules are ambiguous in the way in which certain drawings are ambiguous, that events can make them change face, that, in the heat of battle, the change often remains unnoticed so that we obtain a smooth changeover from one worldview (the paratactic ontology, to take an example) to another, and "incommensurable," one (a more hierarchical arrangement, including a separation of appearance and reality).[32]

Not all conceptual changes are smooth. There is outright destruction (invasions, wars, internal dissension followed by the suppression of the culture and the language of the weaker parties), there is forgetfulness, superficiality, boredom. Many writers assume that the ancient philosophers caused a break of similar magnitude. According to them, the philosophers overwhelmed the naïve worldview of their predecessors and replaced it with a "rational" account. It is of course true that the philosophers, who lived in contentious surroundings, were accustomed to quarrels and debates[33] and had many unusual ideas. What interests me here are ideas that reduced abundance and, among those, ideas resembling the separation of reality and appearance: "realism" is my topic. *Such*

32. We have to say that the structures that preceded the "rise of rationalism" were "open" in the sense that they could be modified without being destroyed. They contained the paths Achilles was about to enter, though in a vague and unfinished way. They were also "closed," for a stimulus was needed to reveal their ambiguities and alternative structures to reset them. Without the stimulus, words, phrases, rules, patterns of behavior would have seemed clear and unproblematic (clarity is the result of routine, not of special insight); without an (existing, or slowly developing) alternative structure, the possibilities implicit in Achilles' language would have lacked in definition. Thus entities such as "geometric perception" or "the archaic form of life" are to a certain extent chimeras; they seem clear when indulged in without much thought; they dissolve when approached from a new direction. The expression "dissolves," too, is somewhat fictitious—the transition often remains unnoticed and amazes or annoys only a thinker who looks at the process from the safe distance of a library, or a book-studded office. As always we must be careful not to interpret fault lines in our theories (recent example from physics: the "fault line" that separates classical terms and quantum terms) as fault lines in the world (molecules do not consist of classical parts and, separated from them, quantum parts). *Ambiguity, however, turns out to be an essential companion of change.* (Cf. "Realism," essay 4 in part II of the present volume).

33. This feature of the Greek public life and its effect on Greek science and philosophy (compared with, say, Babylonian science) is described in G. E. R. Lloyd, *The Revolutions of Wisdom* (Berkeley and Los Angeles: University of California Press, 1987), chap. 2.

*ideas,* I contend, were not invented by a single stroke of the imagination but evolved slowly from a more uniform material. The philosophers accepted them, even froze them with the help of a new instrument, proof, but in doing so they followed history, not reason. I shall now illustrate this process by looking at two early writers, Xenophanes the journalist and Parmenides, whom one might regard as the first Western logician.

# CHAPTER TWO

## XENOPHANES

*1. Character and Circumstances*

XENOPHANES (about 570–470 B.C.) was one of the first Western intellectuals. Like many of his successors he was a conceited bigmouth. Unlike them, he had considerable charm. Although dealing with topics which soon became the exclusive property of a new profession, philosophy, he often behaved in a thoroughly unprofessional way. He used epigrams, one-liners, he imitated, mocked, or repeated popular profundities to reveal their shallowness. Serious thinkers were not amused. Aristotle called him "somewhat uncouth" (*agroiko'teros, Metaphysica* 986b27); modern authors compared him unfavorably with Parmenides. Occasionally they did not even get his jokes.

Some of these jokes are rather obvious. Take fragment B22 (*Die Fragmente der Vorsokratiker*, ed. Diels-Kranz [Zurich: Weidmann, 1985]; my paraphrase):

> A place by the fire in winter is just right for such conversations,
> reclining in softness and comfort and thoroughly stuffed by the meal,
> drinking sweet wine, and nibbling at peanuts and candy:
> "Who are you? From whom? And how many years have you lived?
> And what was your age when the Mede made his appearance?"

The fragment describes a standard situation of the epic: the hero has returned home, he enjoys companionship, wine, a good meal while remembering past dangers. "But are the verses really and entirely serious?" asks E. Heitsch. "Certainly, the contrast between a comfortable present and past suffering is traditional. But the contrast arises here in a very drastic manner. There is nothing in the expressions about eating and drinking that compares with the noble epic formula *ti's po'then eis andron*. On the contrary: in the epic it is the stomach of a slaughtered pig filled with blood and lard and ready to be turned into a sausage that is 'stuffed' in the sense ascribed to the satisfied drinkers here (*Od.* 18.119; 20.26). It is hardly an accident that Xenophanes uses the same word . . . But then his verse sounds slightly amused . . . And this would fit the man who was so critical of all convention."[1]

In B7 Xenophanes ridicules the idea of transmigration:

> Once as he passed by a dog being beaten [he says about Pythagoras] his
>     heart filled with pity
> and it is said that he uttered the following words:
> "Cease and desist and don't beat him—it's the soul of a friend
> That is howling, for I most distinctly can hear his voice."

K. Reinhardt comments as follows: "The barking of the dog as the voice of a well-known friend, the deeply felt sympathy, and the whole so sparingly and yet so securely caricaturized dignity of the great miracle man [Pythagoras] cannot be improved upon."[2]

But there are cases where the mockery seems to hide behind a straightforward factual statement. Take B28:

> This upper end, here, of the earth, lies visibly at our feet.
> touching the air; but the lower continues forever . . .

For F. M. Cornford the fragment is a theory which Xenophanes opposed to the theories of Thales and Anaximenes (who let the earth rest on water and on air respectively).[3] Already Aristotle (*De caelo* 294a21 ff.) accused Xenophanes of trying "to save [himself] the trouble of looking for a reason [for the stability of the earth]." Kirk and Raven call the view "naïve but understandable,"[4] and W. K. C. Guthrie writes as follows: "some of the views described [by Xenophanes] are simply taken over from the Milesians and others are rather nonsensical."[5] The situation looks different when we interpret B28 as a (perhaps semiserious) extension of a novel use of infinity (or indefiniteness) from time to space. The

1. *Xenophanes: Die Fragmente* (Munich and Zurich: Artemis Verlag, 1983), 142 ff.

2. Reinhardt, *Parmenides und die Geschichte der griechischen Philosophie* (Frankfurt: Vittorio Klostermann, 1959), 141.

3. F. M. Cornford, *Principium Sapientiae* (Cambridge: Cambridge University Press, 1952), 147 n. l.

4. G. S. Kirk and J. E. Raven, *The Presocratic Philosophers* (Cambridge: Cambridge University Press, 1960), 176.

5. W. K. C. Guthrie, *A History of Greek Philosophy* (Cambridge : Cambridge University Press, 1962), 1:390.

Milesians had floated the earth on some other substance and had remained silent about the rest; Anaximander had replaced the origin of the world with an in(de)finite past. "If you can eliminate origins by declaring the *apeiron* to be *indefinitely old,*" Xenophanes seems to be saying, "then why not eliminate foundations and the problems to which they lead by declaring the earth to be *indefinitely deep?*" ("goes into the *apeiron*" is a more literal translation of the end of the second line of B28). Seen in this way B28 is anything but simpleminded.

It is his peculiar style, his "lack of professionalism," as a modern critic would say, that explains why serious thinkers hesitate to welcome Xenophanes into their midst. I already mentioned that Aristotle called him "somewhat uncouth" (*agroiko'teros*—*Metaphysica* 986b27—cf. b22: "he made nothing clear") and advised his readers to forget about him. Later authors habitually placed him next to Parmenides as "the lesser figure."[6] For Guthrie, Xenophanes "was in fact a thinker of far less sophistication than Melissus."[7] "In his explanations of the world," writes Karl Reinhardt, "Xenophanes proves to be a philosophical dilettante, keeping everywhere to what is next best and crudest, nowhere capable of following a problem to a deeper level."[8] The learned gentlemen overlook the fact that the idea of professionalism they defend was not available when Xenophanes started writing. Besides, and this they do know, Xenophanes, who lived to be a hundred years old,[9] could be both flippant and serious, even pompous (and then he was duly praised by the professors, as we shall see)—and he belonged to a period of transition. He practiced an old profession—the profession of the teacher/entertainer/reciter who presents and explains the work of the poets—but he also undermined it and, with it, the ideas it tried to preserve.

Homer and the older historians had written in an impersonal way. Hecataeus started his work by mentioning his name and continued by correcting his predecessors. Archilochus, most charming and most

---

6. H. Fränkel, *Wege und Formen frühgriechischen Denkens,* 3d ed. (Munich: C. H. Beck'sche Verlagsbuchhandlung, 1968), 186.

7. Guthrie, *History of Greek Philosophy,* 1:370.

8. Reinhardt, *Parmenides und die Geschichte,* 28.

9. Cf. B8: "Seventy-six years already my thinking drove me through Hellenic lands and, earlier, twenty-five years since my birth—if I correctly tell the truth on these matters."

wicked critic of traditional habits, goes further. Bad is what seems ridiculous to him:

> May a Saian enjoy the shield I left behind in a bush;
> Good protection it was and I regretted to lose it.
> My life I could save—who cares for the shield?
> I'll purchase another—and it will do just as well.
>
> Ernestus Diehl, *Anthologia Lyrica Graeca*, vol. 1
> (Leipzig: Teubner, 1954), fragment 6

To lose one's weapon was shameful for a Homeric Warrior. For Archilochus a shield is a commodity, unconnected with character or morals. Plutarch reports *(Spartan Institutions):* "When the poet Archilochus visited Sparta, he was driven out of the city at a moment's notice, because they discovered [what he had said in the above poem]."[10]
He ridicules the Homeric idea of a hero:

> I can't stand the towering captain strutting along,
> nor him who is waving his locks, not him either who shaves himself well.
> Much I prefer a short one, even with bow legs
> If only he does not retreat and shows grit and courage.
>
> Diehl, fragment 60

This slow adaptation of "thought"—poetry in this case—to the realities of the time was accompanied and accelerated by political events (the Persian conquest of Ionia in 546–44 B.C. is an example), technological wonders (the tunnel of Eupalinos, the bridges of Mandrokles, and the Artemision of Ephesos, a giant structure erected on marshy ground), increased commerce and traffic, and the knowledge gained therefrom.[11] Xenophanes welcomed the developments and saw in them a sign of increasing human independence. Not the Gods, but architects, city planners, shipbuilders, navigators were changing the world. He himself changed his ideas accordingly.

The fragments contain his impressions, his reactions, and, occasionally, even theories of sorts. They cover topics which today are assigned to different subjects. I start with Xenophanes' social criticism.

---

10. Quoted in J. M. Edmonds, *Greek Elegy and Jambus* (London: William Heinemann, 1968), 2:101.

11. For details and references cf. George Sarton, *A History of Science* (1952; reprint, New York: John Wiley and Sons, 1970), vol. 1, chap. 7.

*2. Social Criticism*     XENOPHANES "DARED, he, a Greek of the sixth century, to reject traditional tales as old inventions!" writes Hermann Fränkel.[12] The fragment Fränkel refers to (fragment B1 in the numbering of Diels-Kranz) sounds about as follows:

Clean is the floor, clean are the hands and the cups; and the garlands
freshly now woven, are put on the heads by the boy.
Redolent balsam preserved in the phial is brought by another,
exquisite pleasure lies waiting for us in the bowl;
and a different wine, with the promise not ever to bring disappointment,
soft tasting and sweet to the smell, stands here in the jar.
And in the center the incense dispenses the holy perfume;
cool water is there, full of sweetness and clear to the eye.
Behold the goldyellow loaves and, on the magnificent tables,
overflowing abundance of cheese and rich honey.
And in the center an altar fully covered with flowers
and festive songs sounding all over the house.
But first it is proper for well-disposed men to the God to pay tribute
with words which are pure and stories that fit the occasion;
then, after the common libations and the prayer for strength to act wisely
(the most important concern, preceding all others)
it is not hubris to fill the body with drink—provided
only the old ones need later a slave to get home.
And I praise the man who, having imbibed, can still remember
how much he achieved and how he followed the virtues.
Let him not tell us of battles conducted by Titans and Giants
or even Centaurs—the fantasies of our fathers;
or of civic dissension—not useful are these events.
But one should always pay respect to the Gods.[13]

This poem has various interesting features. First, the surroundings: it is a somewhat restrained party where one thinks of the Gods and does not drink to excess. While some poets, like Alkaeus, praised drinking for its own sake; while those who imitated the Lydians "were so corrupted that some of them, being drunk, saw neither the rising nor the setting of

12. Fränkel, *Wege und Formen*, 341.

13. This is an original translation by Paul Feyerabend. (All translations from the Greek in the present volume are by Feyerabend, except where indicated otherwise.) *Ed.*

the sun" (Athenaeus's paraphrase of the end of fragment 3): Xenophanes advises his drinking companions to drink with moderation so that only the elderly will need a slave to get home. We owe the fragment to precisely these observations: the Sophist Athenaeus of Naukratis quoted it in his main work, an after-dinner conversation on a variety of topics, eating habits included.

A second interesting feature is the conversations of the participants. They do not talk about wars or monsters or political upheavals. They talk about what is "useful" *(chresto'n)*—a new criterion which at that time is found only in Aeschylus (*Persians* 228) and, later, in Herodotus. Personal experiences are important—"how much [they] achieved and how [they] followed the virtues." According to Xenophanes useful matters are furthered neither by Homer (who even in democratic Athens was still the basis of formal education)[14] nor by the modern craze for athletics:

Let him be swift on his feet and in this way defeat all the others;
let him excel in five ways in the grove of the God
here in Olympia, close to Pisanian waters; let him
wrestle, or master the painful profession of boxing
or the terrible contest, known to all as Pankration—
greater would be his honor in the eyes of his neighbors.
Excellent seats would be given to him at the fights and the games
he could eat what he wanted and eat it at public expense;
in gifts they would drown him and permanent property would be his due
and this also if he had proved his ways with a horse,
he, who is lower than I. For my wisdom is better by far
than the brute power of men and of swiftfooted horses.
No, the custom that puts rugged strength over useful achievements
is without sense and should not be further encouraged.
Small is the gain for the city that harbors an excellent boxer
or a fivefold contestant or winner in wrestling,
or an excellent runner who, among all professions

14. T. B. L. Webster, *Athenian Culture and Society* (Berkeley and Los Angeles: University of California Press, 1973), chap. 3. The idea that the older stories are "mere fantasies" occurs already in Hesiod, *Theogony* 27 and in Plutarch, *Solon* 29. Warlike subjects were discouraged by Stesichoros. What Xenophanes adds is that tales of political dissension encourage conceit, which is harmful for the city, and that the "fantasies of our fathers" have no practical use whatsoever.

engaged in competing, is by far the most praised.
Brief is the pleasure the city derives from a contest in Pisa
for it does not fill the stores of the town.

"The greedy way in which these men [the athletes] ate does not surprise us," wrote Athenaeus, who also preserved this fragment. "All participants in the games were invited to eat a lot and also to exercise a lot." To set them up as examples and to admire them is of no use to the city, says Xenophanes.

Nor is luxury, another modern vice. (Xenophanes clearly does not defend all aspects of modernity.)

Having adopted sumptuous Lydian habits,
while still unencumbered by hateful tyrannical rule,
they went to the town place entirely covered in purple—
no less than a thousand in all—
showing proudly their richly adorned coiffure
and dripping with smells well-prepared from exquisite ointments.

*3. Knowledge and Nature*

LATE GEOMETRIC KNOWLEDGE—if we can use such an abstraction—had two properties. It was an enumeration of relatively isolated "facts." This was connected with the "paratactic" character of language which I sketched in section 2 of chapter 1. Second, each single piece of information was acquired in a fairly direct way, by taking a look, following an immediate impression, or considering the structure of an arrangement. Even mathematical proofs which tried to introduce a certain amount of coherence consisted in drawing diagrams or creating patterns that revealed as yet unsuspected properties of numbers, lines, figures. Figure 1, which shows that the sum of the angles in a triangle is the "straight angle," is an example.[15] Divine knowledge was not of a different kind, it was only more comprehensive than human knowledge. It was about "more facts." Xenophanes seems to agree. Complete knowledge, he says, is not given to humans—no human being "has seen" it (B34.1). This con-

15. Cf. A. Szabo, *Anfänge der griechischen Mathematik* (Budapest: Akadémiai Kiadó, 1969), 243 ff. The Pythagorean pebble patterns "showed" some rather abstract relations between integers.

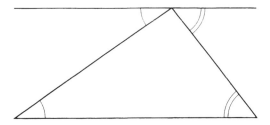

Fig. 1.   The sum of the angles in a triangle is the "straight angle."

tinues the intuitive conception of knowledge. Even if someone succeeded in pronouncing "what has been completed," he continues, he
himself would not be aware of it (B34.3 ff.). Information still seems to be
enumerative and full knowledge (or "what is true," as many interpreters
translate the phrase) an enumeration that has reached its end. Fragment
B34 says that *knowledge* in this sense is inaccessible to humans and that
they are using *conjectures* instead (B34.4).

But why is knowledge inaccessible to humans and what do they get
out of the conjectures?

Considering the Homeric terminology (*tetelesme'non*—meaning
"what has been completed") one might assume that it is the complexity
of things that prevents humans from finding truth. We know fragments;
a complete account, a full enumeration is beyond our abilities. This
sounds plausible until we consider fragment B38: "If God had not created yellow honey, they would believe that figs are much sweeter."[16]

According to this fragment, properties "they" assign to an object depend on circumstances (availability of other objects, their effect on our
sense organs, our judgment, etc.) that have nothing to do with the object, and so the properties that we ascribe to it are therefore not intrinsic
properties of the object. In other words, the fragment suggests that we
distinguish between what an object is, independently of our contact with
it, and what we ascribe to it on the basis of the usual ways of gaining information. This parallels a general tendency: emotionally charged words
such as the "heat" of anger lose their emotional component when being
applied to matter. It also agrees with the habit of physicians like Alkmaion of Kroton, Xenophanes' younger contemporary who separated
symptoms from the illness they were symptoms of and used the word *tek-*

16. The idea is that experiencing the sweetness of honey spoils our taste and actually
changes our experience of figs, which thereafter do not taste as sweet as they used to. *Ed.*

*mairesthai* to indicate how observable symptoms might lead to a not directly observable illness. In both cases the access to a thing gets detached from the thing itself. Being provided with the former but making the latter a criterion of truth we come to the result announced at the end of B34, namely, that "everything is affected by seeming," as Diels-Kranz translates the passage.

Replacing knowledge by conjectures created problems for interpreters such as Fränkel[17] and Reinhardt[18] who did not know how to reconcile Xenophanes' "active realism" (Fränkel), "his very strong preference for reality in every sense: experience, concrete inspection, detail, reasonableness, utility" (Reinhardt) with "even a very tender skepticism" (Reinhardt).[19] The solution is that Xenophanes replaces a criterion[20] which, according to him, turns human opinions into mere seeming, with a new criterion that explains their function in the city.

According to this new criterion opinions are not useless. Just as humans can build, use, and improve ships, bridges, tunnels, in the same way they can introduce, use, and improve another human product, ordinary human utterances (B18 and 35). Like houses, hammers, shovels, such utterances have a function—they provide information. The information may be vague, misleading, and unreliable; but it can be made precise, to the point, and more trustworthy—it can be improved. In his comments on nature Xenophanes tries to improve cosmology by eliminating the "ancestral fables" contained in it.

It is this procedure that gave him the reputation of a crude and unsophisticated thinker. "His theory," says Fränkel, "is strangely primitive and violent even for his own time. It is badly thought out and rather poor."[21] A closer look shows that the primitivism is on the side of the commentators.

In the course of his spring cleaning, Xenophanes eliminated the solid heavens of Homer and the Near East. Reinhardt's comment: "Here Xenophanes did what no Greek philosopher after Thales was permit-

17. Fränkel, *Wege und Formen*, 342 ff.

18. Reinhardt, *Parmenides und die Geschichte*, 151 f.

19. In particular his skepticism of ideas about nature, such as the spherical form of the heavens and of the world, which Xenophanes considered to be part of the ancestral fables and which he replaced by a flat (indefinitely deep) world using unusual conjectures to explain the sun and the rainbow (see below). *Ed.*

20. I.e., a concept of knowledge. *Ed.*

21. Fränkel, *Wege und Formen*, 340. See also 339.

ted to do: he managed to deny the spherical form of the heavens and of the world and with this the connection between the celestial movements."[22]

Now, first of all, it is rather doubtful that Thales assumed any solid celestial sphere. Second, Xenophanes had good reasons. Like the luminiferous ether of the nineteenth century, the solid heavens were part of the "ancestral fantasies" he wanted to remove.[23] Finally, we know now that there is no solid "connection between the celestial movements," that the heavens are open and the spheres nonexistent. Xenophanes was on the right way.

Humans, says Xenophanes, live at the intersection of earth and air and are born from earth and water (B33). Water is the source of rivers, rainfalls, clouds. It seems that Xenophanes found petrified fish and plants in Syracuse, Malta, and Paros and that he regarded them and the water dripping inside caves as evidence for periodic floods. The sun, just like clouds, is new every day, it moves across the surface of the earth and disappears in the West. It is one of the more regular meteorological phenomena. It does not "rise," it does not "set"—such expressions are remnants from the mystical past, only weakly supported by perception (the reader is invited to follow the course of the sun on a cloudy day and to watch its distortions on the horizon). Each country has its own clouds, its own weather, and its own sun.

It is easy to ridicule such a position from our own "advanced" point of view. But the same "advanced" point of view has shown great respect for Parmenides' absurd denial of change and difference, which is contradicted by experiences of the most decisive kind. Reinhardt criticized the removal of the heavens though, seen "from our advanced point of view," this was a sound move. So was the replacement of mythical Iris (the rainbow) by a cloud (B32). The cloudification of the sun is just one more step in the same direction. Moreover, it is not true that Xenophanes "saved himself the trouble of looking for a reason," as Aristotle said, and just blindly produced one primitive conjecture after another. His conjectures are inspired by his new views on knowledge and thus as thoroughly based on "reason" as Aris-

22. Ibid., 146.
23. For the connection cf. W. K. C. Guthrie, *The Greeks and Their Gods* (London: Methuen and Co., 1950), chap. 8.

totle's. So, instead of laughing about the cloudy sun we should admire the attempt to construct a worldview entirely based on experience.

I now come to those ideas that are looked at with approval and are praised for their purity and their critical acumen. Having separated opinion from knowledge and having installed opinion as the only bearer of information, Xenophanes had constructed an early uniform epistemology. His ideas on God destroy this uniformity and introduce what later became the distinction between appearance and reality.

*4. Gods*     XENOPHANES' "THEOLOGY" CONSISTS of a negative part which continues his critique of past prejudices and present follies and of positive statements concerning the nature of divine beings. Some authors assume that he established these statements in a special way, by means of an instrument, later called proof, which made arguments independent of the prejudices or the goodwill of the audience.

Xenophanes' critique of traditional ideas is contained in the following fragments.

> Everything humans despise and condemn and try to avoid,
> theft and adultery and lying deception of others
> Homer and Hesiod laid on the Gods.

B11, 12

> But the mortals consider that Gods were created by birth
> that they wore clothes, had voices, and also a form.
> But if cattle, or lions, or horses had hands, just like humans;
> if they could paint with their hands, and draw and thus create pictures
> then the horses in drawing their Gods would draw horses; and cattle
> would give us pictures and statues of cattle; and therefore
> each would picture the Gods to resemble their constitution.

> Ethiopian Gods—snub-nosed and black
> Thracians—blue-eyed and blond . . .

B14, 15, 16

Here is what some modern writers have said about these lines. Guthrie speaks of "destructive criticism."[24] Mircea Eliade, otherwise an

---

24. Guthrie, *The Greeks and Their Gods*, 1:370.

intelligent gentleman, praises "Xenophanes' acute criticism."[25] Popper reads the fragments as "the discovery that the Greek stories about the Gods cannot be taken seriously because they represent the Gods as human beings,"[26] while von Fritz reads the lines and Xenophanes' positive ideas as expressing a "higher and purer idea of God."[27]

Now it is true that Xenophanes' comments sound excellent when read by a modern progressive intellectual. But that was not its purpose. Xenophanes addressed his contemporaries, not Sir Karl. How did the contemporaries react and what could the defenders of tradition have said in return?

It is not easy to answer this question. Popular religion retained its influence for a considerable time.[28] Xenophanes' abstract divinity (see below) would have given little comfort to those who expected a specific response to specific rituals. Besides, popular religion was even more divided than the Homeric version—different towns had different divinities, even different editions of Zeus who took care of their different ways of life. New Gods were readily added to the existing furniture without any attempt at synthesis or any effort to remove what more narrow-

25. M. Eliade, *Geschichte der religiösen Ideen* (Freiburg: Herder, 1979), 2:407.

26. K. Popper, *Auf der Suche nach einer besseren Welt* (Munich: Piper, 1984), 218.

27. Kurt von Fritz, *Grundprobleme der Geschichte der antiken Wissenschaft* (Berlin and New York: Walter de Gruyter, 1971), 36.

28. The ancient Gods survived until the late Roman Empire. Cf. Robin Lane Fox, *Pagans and Christians* (New York: Norton, 1987), pt. 1, sec. 4, "Seeing the Gods." In Greece the old beliefs were carried on by the leading tragedians.

Aeschylus (who, according to G. Calogero, *Studien über den Eleatismus* [Darmstadt: Wissenschaftliche Buchgesellschaft, 1970], 293 n. 16, was "strongly influenced" by Xenophanes) on the one hand gave the Gods greater and more spiritual power and thereby made them less human; on the other hand he let them participate in the activities of the city (Athena, in the last part of the *Oresteia*, presides over a council of Athenian citizens and casts her vote side by side with them) and so brought them closer to human concerns. The Gods of Aeschylus acted less arbitrarily and more responsibly than the Homeric Gods, arbitrariness and responsibility being measured by the standards of the city. And, of course, the Gods of Aeschylus still were the old Gods, there were many of them, and not just the one Xenophanian power monster.

Sophocles then revived the arbitrariness of the Homeric Gods. Trying to explain the seemingly irrational way in which good and bad fortune are distributed among humans, he attributed both to the actions of equally willful and irrational Gods (see for example *Electra* 558 ff.). Herodotus, whose sentence construction *(lexis eiromene)* and whose tolerance toward conflicting versions of the same story already formally reflects the aggregate view, supported the existence of divine influence with empirical arguments. Thus Xenophanes' arguments were neither valid nor universally accepted. They were propaganda and they affected a narrowly defined group of people.

minded contemporaries might have regarded as contradictions. Thus a fervent defender of religious pluralism could easily reply that, being tribal entities, Gods, like kings, did indeed resemble their subject. "You are right Xenophanes," (s)he might have said. "Our Gods look like us and often act like us. After all, they are our Gods. But what on Earth made you think that that is a criticism?"

On the other hand we have the presumption, also present in certain parts of popular religion, that local Gods are somehow superior to others. The general tendency toward abstraction (which I sketched in section 4 of chapter 1) strengthened this element while weakening the more specific differences. Gilbert Murray ascribes the process to wars and travels. Missing the Gods that were tied to regions they had left behind the Greeks found similar Gods elsewhere and "recognized" them, i.e., they emphasized the similarities over the differences. This extended the reach (= power) of the Gods but diminished their humanity: being "fused or renamed like this, [the Gods became] less living and definite."[29] General notions slowly overwhelmed the more specific images of earlier periods.

This becomes very clear from Xenophanes' positive ideas about God, or from his "theology":

One God alone is the greatest, the greatest of Gods and of men
not resembling the mortals, neither in shape nor in insight.

Always without any movement he remains in a single location
for it would be unseemly to walk now to this, now to that place.

Totally vision, totally knowledge, totally hearing.

But without effort, by insight alone, he moves all that is.

B23, 26, 24, 25

It is interesting to trace the effect this doctrine had in antiquity. We have quotations of key phrases in Aeschylus[30] and we have a comment by Timon of Phleios, a pupil of Pyrrho the skeptic (A35, Diels-Kranz). Timon writes:

29. Gilbert Murray, *The Rise of the Greek Epic* (Oxford: Oxford University Press, 1934), 70.
30. Cf. the first appendix to Calogero.

> Xenophanes, semipretentious, made mincemeat of Homer's deceptions,
> fashioned a God, far from human, equal in all his relations,
> lacking in pain and in motion, and better at thinking than thought.

"Far from human" Timon calls the God of Xenophanes—and he (she? it?) is indeed inhuman not in the sense that anthropomorphism has been left behind but in the entirely different sense that certain *human* properties, such as thought, or vision, or hearing, or planning, are monstrously increased while other, balancing features, such as tolerance, or sympathy, or pain, have been removed. "Always, without any movement, he remains in a single location"—like a well-established king, or high dignitary for whom it would indeed be "unseemly to walk now to this, now to that place." What we have is not a being that transcends humanity (and should therefore be admired?) but a *monster* considerably more terrible than the slightly immoral Homeric Gods could ever aspire to be. These one could still understand; one could speak to them, try to influence them, one could even cheat them here and there, one could prevent undesirable actions on their part by means of prayers, offerings, arguments. There existed personal relations between the Homeric Gods and the world they guided (and often disturbed). The God of Xenophanes *who has still human features,* but enlarged in a grotesque manner, does not permit such relations. Still, he had his effects. Olympianism in its more philosophical and, therefore, moralized form tended to become a religion of fear, a tendency which is reflected in the religious vocabulary. There is no word for "God fearing" in the *Iliad*.[31]

It is strange and somewhat frightening to see with what enthusiasm many intellectuals, then and now, have embraced this and other monsters, regarding them as first steps toward a "more sublime" interpretation of the ground of Being. They should not be blamed for it, however. The idea was "in the air." Only a very strong and emotionally articulated commitment to traditional ways of living could have evaded them. The common people, in the rural areas especially, had such a commitment. Intellectuals, who were city people, who looked down on conventional habits, and whose connection with the lower strata of humanity was never close, lacked it. They lacked the ability to preserve the abundance they and their contemporaries had been entrusted with.

Xenophanes not only *stated* what a proper divinity would be like, ac-

31. E. R. Dodds, *The Greeks and the Irrational* (Boston: Beacon Press, 1957), 35.

cording to some authors he also *proved* at least part of his assertion. But what is a proof, what were its early properties and how did proofs obtain their eminence?

5. *Proof*        THE BEST WAY OF DESCRIBING a proof is to say that it is a story that has special properties. The Homeric epics contain many stores, some long, some concise, all of them interesting. They were accepted because of the power of the tradition to which they belonged. At first sight a proof seems to be a bit of magic; it seems to carry its authority within itself. We hear the story and we accept it—because of its inherent power.

There are some interesting similarities between a proof and a tragedy as interpreted by Aristotle, Corneille, and Lessing. The end of a tragedy, says Aristotle (*De poetica* 7.5) "is that which is inevitably, or as a rule the natural result of something else" which implies (8.4) that "the incidents [of the plot] must be so arranged that if one of them be transposed or removed, the unity of the whole is dislocated and destroyed." Omit "or as a rule" in the first quotation, and you have the relation of the things proved to what goes on before. "What does a poet do who finds in history the story of a woman [Medea] who murders her husband and her sons?" asks Lessing (*Hamburgische Dramaturgie,* sec. 32). And he replies:

> If he is a real poet then his problem will be, above all, to invent a series of causes and effects according to which those improbable crimes *simply must occur* (my emphasis).

Note how real events with their colorful details and their accidental ramifications are reconstructed to fit into a preconceived scheme. The association of knowledge and proof led to similar changes (ambiguous terms being replaced by unambiguous caricatures of them, etc.) and had a similarly compelling effect. It was soon used—and is still being used—to provide "objective" solutions for the problem of cultural plurality, which is as pressing today as it was for ancient philosophers.[32] It will soon emerge that proofs do not solve the problem—they are just as culture dependent as myths and fairy tales: not everybody accepts the reconstructions (of language, for example) that are needed if proofs are to succeed.

32. Other solutions of the problem are opportunism, relativism, and dogmatism. Cf. Kurt von Fritz, "Der gemeinsame Ursprung der Geschichtsschreibung und der exakten

Now consider the following story, which is found in the essay *On Melissus, Xenophanes, and Gorgias* 977a14 ff., which is a product of the Aristotelian school (my paraphrase):[33]

Assume God came into being.

Then either from like, or from unlike.

If from like, then he was already there.

If from unlike, then either from the stronger or from the weaker.

If from the weaker, then the extra strength comes from nothing—but nothing comes from nothing.

If from the stronger, then it is not God.

Hence,

God did not come into being.

Notice the following features of this rather arid story.

It has the form "if . . . , then ————." Such forms are familiar from Near Eastern law. The Sumeric code of Ur-Nammu (before 2000 B.C.) used the even more complicated form "if . . . , provided that _____, then ————."[34] Conditionals occur in the Eshnumma Code of 1800 B.C., the Code of Hammurabi, Exodus 21–23, the nonapodictic laws of Deuteronomy and in early Greek law (Gortyn, Draco). It is plausible to assume that legal experts also knew how to use modus ponens. Considering the international character of the late Bronze Age in the Near East[35] we can further assume that at least some of the pre-Socratics knew the forms and the corresponding procedures.

---

Wissenschaften bei den Griechen," republished in *Schriften zur Griechischen Logik*, vol. 1 (Stuttgart–Bad Canstatt: Frommann-Holzboog, 1978), 23 ff, esp. 30 f. See also my book *Farewell to Reason* (London and New York: Verso, 1987), 5 ff.

33. This proof is on the eternity of God. Xenophanes also gives a proof on the unity of God (also in *On Melissus, Xenophanes, and Gorgias* 977a14 ff.) The latter proof is given in "Realism and the Historicity of Knowledge," essay 1 in part II of the present volume, *Ed.*

34. S. N. Kramer, *The Sumerians* (Chicago: University of Chicago Press, 1963), 83 ff., as well as J. B. Pritchard, ed., *The Ancient Near East*, vol. 2 (Princeton: Princeton University Press, 1975), sec. 3, "Laws from Mesopotamia." Cf. also the excerpts from legal texts in vol. 1 (Princeton: Princeton University Press, 1969), sec. 5.

35. Cf. W. F. Albright, *From the Stone Age to Christianity* (New York: Doubleday, 1957), chap. 4, especially 209 ff., as well as chap. 1 of *Yahweh and the Gods of Canaan* (New York: Doubleday, 1969). For the internationalism in the arts cf. T. B. L. Webster, *From Mycenae to Homer* (New York: Norton, 1964), chap. 3. For the influence of Near Eastern ideas and forms on early Greece cf. Albright, "Neglected Facts in the Greek Intellectual

A second feature is that the proof uses exhaustive and mutually exclusive alternatives. This feature, too, is closely connected with the practice of law. The application of a particular law depends on conditions which must be checked, one after the other, until they are either exhausted (in which case the law does not apply), or a relevant condition is found. The early philosophical proofs, which considered grand schemes, not legal details, simplified the process by providing complete alternatives: one/many; equal/unequal; etc.[36]

A third and most interesting feature is the direction of the argument—it goes from premise to conclusion back to the negation of the premise. Later on this form was called an indirect proof or a reductio ad absurdum. Indirect proofs were not always welcome and their application was restricted by Brouwer. Indirect arguments are older and apparently arose from the subjective: X might have occurred had Y occurred (the hidden assumption being that Y and, therefore, X did not occur).[37]

Tying these features together into a single story, the author of the proof created a text that is quite compelling. Did the story establish divine eternity in an "objective," i.e., culture-independent way? Not at all! I already mentioned that a proof needs stable and unambiguous concepts. For example, it is no longer possible to see divinities as having now a restricted domain and on another occasion immense power. A second reason is the particular precise concept of divinity that is being used. The proof assumes that being divine means the same as having supreme power. The Homeric Gods did not satisfy this criterion. The succeeding development approached it, as I already mentioned. Hence, if the proof achieved its aim then the reason is that the notion of divinity had already been bent in its direction: *history, not argument, undermined the Gods.*

Revolution," *Proceedings of the American Philosophical Society* 116 (1972): 225 ff. For the social background cf. G. E. Mendenhall, "Ancient Oriental and Biblical Law," *The Biblical Archaeological Reader* 3, ed. Edward F. Campbell and David Noel Freedman (New York: Doubleday, 1970). "Some Comparative Considerations" going beyond the law and the Near East are offered by G. E. R. Lloyd, *Magic, Reason, and Experience* (Cambridge: Cambridge University Press, 1979), chap. 2.

36. G. E. R. Lloyd, *Polarity and Analogy* (Cambridge: Cambridge University Press, 1971), chap. 2, discusses the historical background and some early uses of these simplifications.

37. G. E. R. Lloyd, *Magic, Reason, and Experience* (Cambridge: Cambridge University Press, 1979), chap. 2.

The apparently remorseless power of proof stories made an immense impression when it was first perceived. Plato's *Euthydemus* shows the fascination it held for an Athenian audience, two generations later.

> Logic was the rage of the day; all over, at the marketplaces, in the streets, in private homes and in public buildings, at all times, sometimes all through the night, people engaged in dialectical disputations and flocked to hear the acknowledged masters of logical argument display their art. The writings of this period mirror this development in thought: drama as well as the dialogue form bear witness to the power which the newborn science wielded over the imagination of men. (D. E. Gershenson and D. A. Greenberg, "The Physics of the Eleatics," in *The Natural Philosopher*, vol. 3 [New York: Blaisdell Publishing, 1964], 103)

The Homeric heroes had argued, too. An argument consisted of showing that one's side had the support of tradition. Tradition was applied directly, occasionally in the form of one-liners (e.g., Aeschylus, *Eumenides* 585 ff.). What was missing was a procedure that divided the argument into steps, established strong connections between them, and used alternatives to create an impression of exhaustiveness. It would be interesting to find where and under what circumstances the peculiarities of legal practice, medical argument, protoarithmetic, and protogeometry gave rise to explicit sequences of statements such as the one I presented above. At any rate, the apparent power of such sequences must have been overwhelming when it was first perceived. Making an idea part of a proof meant, of course, changing it, just as making a story part of drama entailed important and not always realistic changes. In the case of the Gods (and, later, of Being) the implied constraints paralleled the constraints of history and accelerated the creation of a new domain over and above experience and tradition, a domain that is now being called reality.

Kurt von Fritz believes that the argument was constructed by Xenophanes.[38] If that is the case then Xenophanes discovered two ways of ascertaining a state of affairs—experience and proof—and he may have considered two corresponding domains of information—phenomena

38. Cf. K. von Fritz, "Xenophanes," in *Realencyclopaedie der classischen Altertumswissenschaft*, ed. August Friedrich Pauly and Georg Wissowa (Stuttgart: Alfred Druckenmüller Verlag, 1972), Reihe 2, Halbband 19. One of his arguments is that the word *homoios* occurs in a strict sense that disappeared later.

and realities. In both cases Xenophanes started from common ideas; in both cases it seemed that he had revealed hitherto unknown properties of their objects. The ambiguity of the ideas from which he started and the developments which carried them along were important preconditions of his success.

CHAPTER THREE

# PARMENIDES AND THE LOGIC

# OF BEING

*1. The Way of Truth*    PRESENTED BLUNTLY and without preparation, Parmenides' idea that change and
difference are illusions seems to be "next
door to lunacy" (Aristotle, *De generatione et corruptione* 325a18 f.). The
idea becomes plausible, however, when compared with preceding and
later points of view.[1]

Like modern scientists, the Ionian cosmologists, Thales and Anaximander among them, posited simple principles to explain the diversity of
events. Modern scientists use laws; the Ionians used substances.[2] Thus
Thales assumed that, basically, everything consists of water. Thales may
have argued for his choice; for example, he may have pointed out that water occurs in a solid, a fluid, and an airy form, that the forms readily
change into each other, and that water is necessary for life. In fact, the
Greeks, who lived around the Mediterranean "like frogs around a pond,"
had firsthand experience of the immensity and the many forms and functions of water. Water turns into mist which, being heated by the sun, turns
into air; alternatively, the mist may condense into clouds which produce
rain, fire (lightning), and solid substances (hail). Thales' successors used
comparable arguments in favor of other substances. Then Anaximander
pointed out that earth, water, air, and fire were equally important and that
the basic substance, therefore, must be different from all of them. We can
understand Parmenides when assuming that he pushed this approach still
further, using the new instrument of proof to harden his assumptions.
What underlies nature, he may have said, must indeed be different from
nature and more comprehensive. According to Parmenides the most basic entity underlying everything there is, including Gods, fleas, dogs, and

1. Cf. also my *Farewell to Reason* (London and New York: Verso, 1987), 61–70, 120–22.
2. Cf. also my *Against Method* (London: Verso, 1987), 42 (p. 43 in the 1993 ed.).

60

any hypothetical substance one might propose, is Being. This was in a sense a very trivial but also a rather shrewd suggestion, for Being is the place where logic and existence meet: every statement involving the word "is" is also a statement about the essence of things.

Having made Being his basic substance, Parmenides considered the consequences. They are that Being is *(estin)* and that not-Being is not. What happens on the basic level? Nothing. The only possible change of Being is into not-Being, not-Being does not exist, hence there is no change. What is the structure of Being? It is full, continuous, without subdivisions. Any subdivision would be between Being and something else, the only something else on the basic level is not-Being, not-Being does not exist, hence there are no subdivisions. But is it not true that we traditionally assume and personally experience change and difference? Yes, we do. Which shows, according to Parmenides, that neither tradition nor experience provides reliable knowledge. This was the so far clearest and most radical separation of the domains which later on were called "reality" and "appearance." It was also the first and the most concise theory of knowledge. Theories of knowledge try to explain how familiarity with one domain (perception, for example) leads to knowledge about another that is independent of it (reality). Parmenides answers that this never happens, that Being must be approached directly, that the one agency that can approach it directly is reason, that revelation taught him, Parmenides, how to use reason, and that he is now capable of explaining this use to others.

Those who like to make fun of Parmenides and who call his arguments primitive and linguistically absurd should consider how many modern scientists repeat his general ideas without his rigor and coherence. To start with, the premise, *estin*—Being is—is the first explicit conservation law; it states the conservation of Being. Used in the form that nothing comes from nothing (which found its way into poetry: *King Lear* 1.1.90) or, in Latin, *ex nihilo ni(hi)l fit*, it suggested more specific conservation laws such as the conservation of matter (Lavoisier) and the conservation of energy (R. Mayer, who begins a decisive paper with this very principle).[3] Nineteenth-century point mechanics posited a "real"

---

3. "Die organische Bewegung in ihrem Zusammenhang mit dem Stoffwechsel," in Robert Mayer, *Die Mechanik der Wärme*, Oswalds Klassiker der exakten Wissenschaften, no. 180 (Leipzig: Verlag Wilhelm Engelmann, 1911), 9.

world without colors, smells, etc., and a minimum of change; all that happens is that certain configurations move reversibly from one moment to another. In a relativistic world even these events are laid out in advance. Here the world

> simply *is*, it does not *happen*. Only to the gaze of my consciousness, crawling upward along the lifeline of my body, does a section of this world come to life as a fleeting image in space which continuously changes in time. (H. Weyl, *Philosophy of Mathematics and Natural Science* [Princeton: Princeton University Press, 1949], 116)

"For us," wrote Einstein,

> who are convinced physicists, the distinction between past, present, and future has no other meaning than that of an illusion, though a tenacious one. (*Correspondance avec Michèle Besso*, ed. P. Speziali [Paris: Hermann, 1979], 312; cf. 292)

Irreversibility, accordingly, was ascribed to the observer, not to nature herself. The trouble is that "convinced physicists" are also "convinced" empiricists. But how can experiments which occur in time and are therefore illusions tell us about a reality that is beyond all illusions?

Max Planck recognized the problem but did not solve it. His essay "Positivismus und reale Aussenwelt," which he first read in 1930 contains the following passage:

> The two statements, "There exists a real external world which is independent of us" and "This world cannot be known immediately" together form the basis of all of physics. However, they are in conflict to a certain extent and thereby reveal the irrational element inherent in physics and in every other science, which is responsible for the fact that a science can never solve its task completely. (*Vorträge und Erinnerungen* [Darmstadt: Wissenschaftliche Buchhandlung, 1945], 235 f.)

Moreover, the "immediate sense impressions" which Planck, Einstein, and other empiricists regard as a foundation of knowledge[4] are not part

---

4. A. Einstein, letter to M. Solovine of May 1952, in *Einstein: A Centenary Volume,* ed. A. P. French (Cambridge: Cambridge University Press, 1979), 270; Planck, quoted ibid. ("there is no other source of knowledge than our sense impressions"). According to Einstein, "Physics and Reality," quoted in *Ideas and Opinions by Albert Einstein* (New York: Crown Publishers, 1954), 291 f., we start from a "labyrinth of sense impressions," select from it "mentally, and arbitrarily, certain repeated occurring complexes of sense

of our experience (which is an experience of objects in space) but the-oretical constructs that have to be unearthed by special methods (re-duction screen, etc.). Thus we have here a view in which a hidden re-ality thoroughly independent of human events is said to be based on hidden processes extremely dependent on them. One cannot say that things have improved since Parmenides. And it is perhaps not entirely useless to return to him and to examine the reasons he gives for his po-sition.

Parmenides distinguishes between common beliefs and the actual course of the world. And he uses the newly found instrument, proof, to immobilize his concepts and to establish his results.[5] Like Xenophanes he starts from a notion which, using ambiguity, can be found in common sense but which he changes to fit his intentions. "What he tried to show," writes Reinhardt about Xenophanes' proof,[6] "was the unity of God. To do this he chose the concept of supreme power. It did not occur to him or, at least, it did not disturb him that this concept was no more familiar than the other one [viz., the concept of unity]—popular religion was not acquainted with either; for only the concept of unity was accessible to di-alectic treatment—and dialectics was his main concern."[7] The form of the argument affects the premise, the nature of God there and the struc-ture of Being in the present case. Also the elements are now tied together much more strongly. All this does not make the case more "objective." On the contrary, it shows the changes that are needed to make the story convincing—and these changes need not be accepted. Yet the author treats his results as having been established independently of all prefer-

---

impressions," correlate them with the concept of a bodily object and attribute to this con-cept "a significance which is to a high degree independent of the sense impressions which originally give rise to it." Reality, according to this account, is a "mental and arbitrary" construction, introduced to "orient ourselves" in a "labyrinth of sense impressions." Now, first of all, such a "labyrinth" is nowhere to be found in our lives; it is itself an "arbitrary" construction. Second, how can we introduce the order we need to "orient ourselves" while still lacking it, i.e., while being disoriented?

5. For details and further references cf. the articles in R. E. Allen and D. J. Furley, eds., *Studies in Presocratic Philosophy*, vol. 2 (London: Routledge and Kegan Paul, 1975), especially G. E. L. Owen, "Eleatic Questions." Parmenides' influence upon mathemat-ics is discussed in A. Szabo, *Anfänge der griechischen Mathematik* (Budapest: Akadémiai Kiadó, 1969), 287 ff.

6. See chapter 2 note 33. *Ed.*

7. Karl Reinhardt, *Parmenides und die Geschichte der griechischen Philosophie* (Frank-furt: Vittorio Klostermann, 1959), 96.

ences. Parmenides' reasoning, though more concise, is open to the very same objection.

Parmenides tells his story in an epic poem. He could have chosen other forms: lyrical poetry, tragedy (example: the debates and the political ending of the *Eumenides*), scientific prose (a new medium, used by the Ionian philosophers, especially by Anaximander), the lampoon (employed with excellent effect by Archilochus and Xenophanes), and various forms of public oratory. (The dialogue came into prominence only later.) His situation differed from that of modern writers who have no choice but to simply follow professional standards, editorial policies, and stylistic inertia. Parmenides adopted the epic form and the associated mythological inventory. Not prose, but the hexameter is his medium; not he, but a Goddess explains the world, and the nature of knowledge. There are two parts, one dealing with "what is necessary" (Diels-Kranz, *Die Fragmente der Vorsokratiker* [Zurich: Weidmann, 1985], fragment B2.4) or, using later notions, reality, the other with appearance (the beliefs of the multitude). It seems that, like Xenophanes, Parmenides thought that both parts were important—why else would he have developed a story of change which, as far as we can see, was much longer than his exploration of Being? I shall give a rather streamlined account of the poem. The mythical elements which play a far from negligible role[8] and which can be used to support a very different approach will be disregarded. Thus curtailed and paraphrased the "path of truth" may be subdivided into four sections: (1) an announcement concerning the procedure to be followed; (2) the premise used; (3) the program of the proof; and (4) the proof.

(1) According to the announcement (fragment B7.3–5) the key assertions of the story are to be established neither by habit derived from experience *(ethos poly'peiron)*, nor by the aimless eye or the echoing ear, nor even by mere talk; they will be established by proof *(logos)*.

The rejection of experience and traditional wisdom was not new. It inspired various religious movements and had already been articulated in the *Iliad* (see chapter 1). Parmenides sharpens the difference and declares proof—not belief, not religious instruction, not purification of mind and soul—to be the basis of truth.

8. Ibid., 64 ff.

(2) The premise is *estin*—*est* in Latin or, somewhat misleading, *it is* in English. The reader who thinks that nothing can follow from a mere word should consider formulae such as E = const (principle of the conservation of energy), which, combined with further assumptions, yields an enormous amount of detailed information.

(3) "On the way [to the truth] there are many signs" (B8.2). What will be proved is (B8.3 f.) that "what-is" is uncreated and indestructible; whole, and uniform (continuous—*syneche's*); complete and unmoving.

(4) The proof itself consists of interconnected parts, each part being linked to its predecessor by the word *epei*, 'because'.[9]

First it is proved that what is cannot have come into being *(age'netos)*. Why?

> Because it cannot have come from not-Being (the reason is given below, under A)
> Nor can it have come from Being (reason given below, under B)
> But either Being, or not-Being (B8.16) hence,
> what-is cannot have come into Being.

Reason A is that not-Being can be neither described nor known nor said to give rise to anything at any particular time (the event could always have happened at a different time). Reason B is that what-is is and therefore cannot grow.

Note the two features already commented upon: the use of a set of alternatives assumed to be exclusive and complete and the indirect procedure. Both added to knowledge, especially to mathematical knowledge, and made it more abstract. The discovery of incommensurable lengths (based on the alternative even-uneven) was a natural consequence (cf. the proof interpolated in Euclid, *Elements*, 10.117). The idea that what-is-not cannot be talked about turns both the genetic cosmologies and

---

9. For this feature see the "Additional Note A," in Owen, "Eleatic Questions," 76 f. According to Lloyd, *Polarity and Analogy* (Cambridge: Cambridge University Press, 1971), 104 ff., Parmenides' *unaltered existence* and *unaltered nonexistence* are contraries, not contradictories, and the "proof" itself is therefore invalid. "Parmenides forces the issue," says Lloyd (105). I agree though I put the "forcing" in a different place—with what I say below about the assumption implicit in the proof (the proof "assumes that what exists simply is . . . and has no further properties").

their Parmenidean refutation into mere noise—the disproof disappears (becomes meaningless noise) together with the thesis disproved. Note again that *estin* as used by Parmenides is the first conservation law in the history of Western science; it proclaims the conservation of Being.

Next it is proved that what-is cannot be subdivided. The steps are the same as before except that the suppositions now deal with spatial, not with temporal differences.

It is inferred that what-is is completely connected *(syneche's)*. The implied idea of continuity was elaborated by Zeno and Aristotle and mentioned by Weyl (details in my *Farewell to Reason,* chap. 8). Considering that atomism, humoral pathology, the theory of elements on which it relied (and which was still defended by Lavoisier), the split of medicine (and other subjects) into an empirical (clinical) and a theoretical branch, Aristotle's creation of a qualitative physics, and Galileo's reaction to it are all connected with the ideas that underlie Parmenides' poem, we have to admit that his emphasis on the unity of Being was an important ingredient of Western intellectual life. But did Parmenides succeed in proving the case independently of any extra-argumentative tendencies and preferences?

The argument *wants to prove* that "reality" is eternal, indivisible, and free from change. It *assumes* that what exists simply is—*estin*—and has no further properties. Once this assumption is made, the only distinction that remains between an event and its predecessor in time (or neighbor in space) is that the one is and the other is not—and now the conclusion follows. The assumption was perhaps motivated by the general trend toward abstraction I already mentioned on various occasions, by theories (such as Anaximander's cosmology, which Parmenides seems to have known) that articulated the trend, by the requirements of the logical machinery Parmenides used, or by a personal conviction that transcended all these stimuli. The assumption, however, was not established in the way he regarded as the only legitimate one—by reasoning.

*Estin* was a premise and so it certainly was not established by the argument itself. More importantly, there existed objections against accepting such an assumption. Aristotle, in his discussion of Parmenides and Plato (*Physica* 184b25 ff., *De generatione et corruptione* 325a18 ff., *Ethica Nicomachea* 1096b33 ff.), mentions two: the assumption conflicts with natural philosophy (where change and subdivision are taken for granted); and it conflicts with common sense ("'to be' is used in many ways"—a favorite Aristotelian slogan). Having less (social, practical, psychological) power behind it than either of these objections and functioning, as we have seen, as an unsupported starting point and not as a result of reasoning, the assumption has to give way: the stronger tradition overrules the weaker even when the latter is held together by tight logical connections. In the next section I shall comment on the features and the force of such a "social refutation" of a logical point.

To summarize: Parmenides, like Achilles, is driven by strong motives. Like Achilles he can elucidate the object of his motives by analogy to already existing, though vague, conceptions. Unlike Achilles he seems to possess a new way of fortifying ideas—a kind of protologic. The protologic was prepared by linguistic forms (the subjunctive), by a widespread legal practice, and by the refinement these elements experienced in a society devoted to debate and intellectual competition.[10] It was accepted by an ever-increasing number of people and eventually became the "rage of the day." Thus a powerful vision that agreed with a general tendency toward abstraction and could be explicated—but not proved—by a popular instrument of thought assumed a solidity that seemed to make it independent of the accidents of belief and history. Let us see what this situation teaches us about the idea of reality.

*2. Antilogike*     PARMENIDES ARGUED BY MODUS TOLLENS: he made an assumption, developed its consequences, found them to be false, and rejected the assumption. Can one refute him in the same manner? For example, by pointing out that change, which is denied by him, quite obviously exists? In other words— is it possible to restrict the debate to facts and some simple argumentative pattern or do motivations, outside structures and historical tendencies enter all debates? In the case of Achilles I suggested that they do. Can this result be generalized? Does it perhaps only apply to the sciences? Let us see what ancient writers had to say on the matter.

Plato's dialogue *Theaetetus* deals with the nature of knowledge. Various views are proposed, the thesis that knowledge is perception among them (151e2 f.). Socrates articulates the view, quotes authorities such as Homer, Heraclitus, and Protagoras, and describes how, following these

---

10. Cf. G. E. R. Lloyd, *The Revolution of Wisdom* (Berkeley and Los Angeles: University of California Press, 1987), chap. 2.

authorities, perception must be defined (153d8 ff.; 156b6 ff.): neither the object nor the perceptions can exist by itself; act, object, and perception form an indivisible block. The unperceived world contains motions, some swift, some slow, but without definite properties. The motions interact and produce the block with its well-defined sensation. This is a striking anticipation of some features of quantum mechanics (Einstein-Podolsky-Rosen correlations), for here, too, it makes no sense to speak of the spatiotemporal properties of objects except as part of a measurement. Finally Socrates raises various objections. For example, he says that if knowledge really were perception then we would cease to know what is before us when we close our eyes—but we still know because we remember what we have seen (163e4 ff.). Or that closing one eye we would both know and not know (165c4 ff.). Or, he says, that if knowledge were perception, it could be indistinct, blurred, fading, while knowledge does not admit degrees. Or he points out that identifying knowledge with perception makes teachers superfluous, but Protagoras who holds the view still claims to be a teacher (161c1 ff.). "We must say, then," Socrates concludes his diatribe, "that each of the two [viz., knowledge and perception] is different from the other" (164b11). "Obviously," mumbles Theaetetus, his fall guy.

Socrates collects counterexamples to make his point. This was a straightforward and widely used type of argument. It played a large role in Parmenides and in some early mathematicians who constructed their counterexamples instead of merely collecting them (later on their method was called a reductio ad absurdum); it was used with exuberance by the Sophists. Counterexamples are still popular though under a different name (falsification), with a different domain of application (the sciences, mainly) and different father figures behind them. Some writers believe that scientific knowledge can be changed and improved by using invention, conditions of adequacy and content, falsification—and nothing else. Applying the method to Parmenides they explain the rise of atomism in the following way: trying to criticize Parmenides, Democritus (or Leucippus) looked for counterexamples, found change, refuted the view that Being was stable and undivided, and replaced it with something better.[11]

11. An example is K. R. Popper, *Realism and the Aim of Science* (London: Hutchinson Group, 1983), xxvi.

Such an account cannot possibly be correct. It suggests that Parmenides, being overwhelmed by his vision, did not notice change while Democritus, more a man of the world, discovered it and refuted the Parmenidean theory. But Parmenides, far from overlooking change, tried to explain it (in the second part of his poem), though with the restriction that he was dealing with appearances; reality, he said (though not in these words), is unchanging and undivided. Democritus and Leucippus, accordingly, had *three* tasks before them: to *(re)define* reality (and knowledge); to *connect* the newly defined entity with opinions (which regained their status as knowledge or, at least, as useful information); and to *explore* it on that basis.

Aristotle considered all these tasks and made various comments on the first. He criticized Parmenides for deviating from common sense. First, the criticism was *possible:* Parmenides had not proved his case but had started with the deviation (see the preceding section). Second, the criticism was *motivated* by the wish to bring theory closer to the ideas and actions of life in the city. "Even if there existed a Good," writes Aristotle (*Ethica Nicomachea* 1096b33 ff., my emphasis), commenting on analogous tendencies in Plato,

> that is one and can be predicated generally or that exists separately and in and for itself, it would be clear that such a Good can be neither produced nor acquired by human beings. *However, it is just such a Good that we are looking for* . . . one cannot see what use a weaver or a carpenter will have for his own profession from knowing the Good in itself, or how somebody will become a better physician or a better general once "he has had a look at the idea of the Good" [apparently an ironical quotation of a Platonic formula]. It seems that the physician does not try to find health in itself, but the health of human beings or perhaps even the health of the individual. For he heals the individual.

And, third, the criticism was more than an intellectual game—it had *substance.* Even the most abstract thinkers have to consider (and many early philosophers did consider) their duties as citizens. It seems that Democritus also wanted a closer connection between abstract knowledge and common experience. Leucippus, who is traditionally associated either with the Eleatics in general or with Parmenides' disciple Zeno,

> thought he had a theory which harmonized with sense perception and would not abolish either coming-to-be and passing-away or motion and the multi-

plicity of things. He made these concessions to the facts of perception. On the other hand he conceded to the Monists that there could be no motion without a void. The result is a theory which he states as follows: "The void is a 'not-Being,' and no part of 'what is' is a 'not-Being'; for what 'is' in the strict sense of the terms is an absolute *plenum*. This *plenum*, however, is not 'one'; on the contrary, it is a 'many,' infinite in number and invisible owing to the minuteness of their bulk." (Aristotle, *De generatione et corruptione* 325a23 ff., Ross trans.)

In short, Being is many and moves in not-Being. Note the nature of the argument: Leucippus does not try to refute Parmenides by using the fact of motion. Parmenides had been aware of the "fact" and had declared it to be illusory. Moreover, he had not simply asserted the illusory character of motion, he had presented proofs. He had transcended sense impression on the grounds "that 'one ought to follow the argument'" (Aristotle, *De generatione et corruptione* 325a12 f.). Leucippus, in contrast, decided to follow perception; one might say that he and those who thought in a similar manner (Democritus, Empedocles, Anaxagoras) wanted to bring physics closer to common sense.[12]

This was not a matter of course. Religious and philosophical movements both in Greece and elsewhere denigrated common beliefs and customs and tried to lead people away from the trivialities of daily life. Pythagoras and Parmenides had constructed a science and a philosophy of precisely this nature and had supplied them with new means of persuasion. Mystics, seers, shamans all over the world try to prove that an uncommon life is not only possible, but real or, to use "Parmenidean" terms, they claim to have reached and experienced the Parmenidean One.[13] One cannot object that such a state lies outside common experi-

12. The different notions of reality that are implied here, and their consequences for modern science and philosophy, are further developed in "Realism," essay 4 in part II of the present volume. *Ed.*

13. Cf. the excerpts in S. Radakrishnan and Ch. A. Moore, eds., *A Source Book in Indian Philosophy* (Princeton: Princeton University Press, 1957), e.g., 327. For the Pythagoreans, cf. Walter Burkert, *Lore and Science in Ancient Pythagoreanism* (Cambridge: Harvard University Press, 1972), and B. L. van der Waerden, *Die Pythagoreer* (Zurich: Artemis Verlag, 1978).

More recently, the quantum theory invites us to regard the world as a single indivisible whole. The objection that people quite obviously are different from the chairs on which they sit and from the enemies they despise has as much force as the objection that the sun rises and that Copernicus, who says that the horizon falls, cannot be right. In both cases the debate is not only about facts, but also about the role they are supposed to play in our lives.

ence, that it needs preparation, is difficult to achieve, and does not last. The same is true of the observations of the W and Z particles and of the neutrino,[14] all of which are now regarded as "real." What matters is that the state exists, at least approximately, that some people strive for it, that they make it the center of their lives, and that they define reality (in words, or by the way they live) in relation to that center. An opponent must therefore do more than provide facts, rules, and arguments resting on them. He must dismantle the definition and change the life to which it belongs. Arguments about reality have an "existential" component: *we regard those things as real which play an important role in the kind of life we prefer.*

Democritus and Aristotle moved change and subdivision from the periphery toward the center of philosophy, the former a little, the latter to a much larger extent. They moved it as a matter of course, for change was a prominent feature of the lives they led and were not prepared to give up. Speaking in an abstract way we may say that both Democritus and Aristotle made a practical decision.[15] Aristotle occasionally acts as if alternatives were not just undesirable, but absurd.[16] Having abandoned absolute stability while still a member of Plato's school he is now only concerned with developing the consequences of that step.

I shall often return to this "existential" feature of discussions about reality. It was prominent in Achilles—remember that it was his disappointment that made him see a different world—and now returns in more theoretical clothes. For the moment let me say that it undercuts claims such as these: that Parmenides, by the sheer power of his mind

14. P. Watkins, *Story of the W and Z* (Cambridge: Cambridge University Press, 1986).

15. Erwin Schrödinger used precisely such a decision in his criticism of Bohr: "Bohr's standpoint that a spatiotemporal description is impossible, I reject *a limine*. Physics consists not merely of atomic research, science not merely of physics, and life not merely of science. The purpose of atomic research is to fit our experiences from this field into the rest of our thought; but the rest of our thought, as far as it has to do with the external world, moves in space and time." W. Wien, *Aus dem Leben und Wirken eines Physikers* (Leipzig: Barth, 1930), 74. Galileo went the opposite way: the "rest of thought" must be adapted to the ideas of philosophers and astronomers. See below.

16. "Although these opinions appear to follow if one looks at the argument, still to believe them seems to stand next door to lunacy when one considers practice. For in fact no lunatic seems to stand so far outside as to suppose that fire and ice are one." *De generatione et corruptione* 325a ff.

and unmoved by external agencies, discovered the unity of Being; that Democritus and Aristotle refuted him by research, not by merely declaring a well-known process to lie at the center of a preferred way of living; that Copernicus and Galileo proved the motion of the earth without any preceding, associated, or incipient change concerning what is and what is not important in human life; that Darwin similarly proved a connection between humans and other species without any support from changing evaluations, ideologies, social tendencies—and so on. In all these cases we have a change or a tendency for change (which may be implicit or active in special groups only and may be accompanied by contrary tendencies) followed by a theoretical analysis of the products of the change, all of this embedded in an ideology that declares the analysis to be independent of choices, desires, social tendencies and to be a prominent cause of the change. Even very simple arguments exhibit this pattern.

"But what are we doing?" Socrates asks after his diatribe (164c1 ff.). "Like a bad cock we forget about our topic and start crowing long before we have been victorious." The objections, he says, are not really decisive. He also explains why. All the debaters did was "to base [their] objections on the similarity [or dissimilarity] of words" (164c8 ff.): starting from the thesis (that knowledge is perception) and developing it further, always taking care to retain the word "knowledge" or "to know," they arrive at a conclusion that also contains the word, but "seems impossible," namely, that "somebody who has obtained knowledge by remembering [a thing] still doesn't know it because he doesn't see it" (164b4 ff.). What we have is a clash between a sentence that follows from the thesis and contains the word "know" and another sentence (whoever remembers something also knows it) of unexamined origin and authority that contains the same word, but used in accordance with common sense, and therefore seems eminently plausible. Regarding this clash as a refutation of the thesis, says Socrates, is *antilogike,* word bashing; it has nothing to do with the search for wisdom.

Plato uses the word *antilogike* in various places. Its meaning "tends to be whatever Plato thinks of as bad method at the moment."[17] One com-

17. R. Robinson, *Plato's Early Dialectic* (Oxford: Clarendon Press, 1962), 85. Robinson gives places and explanations.

mon element is that opinions are set against opinions, now in one way, now in another (*Phaedrus* 261c4 ff.; *Sophistus* 232b6; *Phaedo* 90c2 ff.). Another is the importance of the *words* that make up the opinions. "Clinging to a word," the controversialist "pursues the opposite of what has been said" and thus conducts only a verbal quarrel (*Republic* 454a1 ff.; cf. *Theaetetus* 164c8 ff.). And, finally, opinions are used as they come, without analysis or proper order (*Phaedo* 101e1 ff.).

The argument in *Theaetetus* 163a1 ff. is a case in point.

Knowledge, says Protagoras, is perception [1]. When I see something I therefore know what I see [2]. When I close my eyes I no longer see it and therefore, according to Protagoras, no longer know it [3].

But I remember it. Hence either I don't know what I remember—which is absurd [4], or I do know it [5] and yet also don't know it because my eyes are closed [6], which is again absurd [7]. So, [1] must be given up.

We start from a thesis and assume we understand it. [2] is an obvious consequence; so is [3], provided nothing else occurs. If I remember the object then [5] and therefore [7] may be generally believed—but the question is: what does Protagoras say about memory? He has an unusual thesis. Would he not also have some unusual assumptions about memory and personal identity? Might he not say, for example (166b1 ff.) that the seen object is not the same as the remembered object and that it is therefore indeed correct to say that, having closed one's eyes, one does not know the object any longer, no matter how many memories there are? Or, taking different organs (such as a closed eye and an open one—165b7 ff.) and even different states of a person (Socrates seeing, Socrates remembering—166b5 ff.), that a person can know and not know at the same time? And might he not reject either [4] or [6] as taking for granted the very things he intends to deny?

Proceeding further along this line of reasoning we must admit that the key words of an argument are often ambiguous in the sense that they await specification from the kind of enterprise one is engaged in. If the purpose is to change beliefs in accordance with a new and comprehensive cosmology, then a conflict between this cosmology and popular opinion cannot be used to criticize the former. If the purpose is to explore the ramifications of a new philosophy, and perhaps to find its limits, then popularity again does not amount to much, for it is to be expected that new views will conflict with "old prejudices" (so called from the standpoint of the new views). If, on the other hand, the purpose is to maintain a certain form of life, then the verbal manifestations of this form of life will become important measures of adequacy. At any rate— the provenience of a sentence containing a key word of an argument has to be checked before the sentence can be used as part of the argument.

For example, the objection from memory assumes that a perception and the corresponding memory both refer to one and the same fact, which becomes known either by perception or from memory. But the theory that knowledge is perception rejects perception-independent and memory-independent facts, and should not be criticized by resurrecting them. The objection that knowledge, being perception, might become indistinct similarly criticizes the theory by showing that it differs from the ways of speaking inherent in an already accepted view. A point of view, Socrates implies, must be permitted to transform beliefs and linguistic habits and should be criticized only after the needed changes have been carried out. How does Socrates argue with such a principle before him?

To start with, he introduces an interesting ambiguity. Knowledge and perception seemed to be clear and definite entities and so seemed the thesis that identified the two. But the identification led to conflict. If we still want to maintain the thesis as Socrates advises us to do then we must change either one entity, or the other, or both. We must change them— but without ceasing to examine the thesis, i.e., without ceasing to look for obstacles. What obstacles? The obstacles that arise after the key terms have received a new content. Socrates provides a new sense for "perception"—the quantum mechanics analogon mentioned above—but not for "knowledge." Does he stop arguing? He does not—he only changes direction. For example he points out (181b8 ff.) that Protagoras leaves no stability and makes knowledge impossible. The remark assumes that knowledge does not participate in the processes Socrates introduced when explicating perception (153d3 ff.). The assumption makes definite what seemed to have become vague, but as part of the criticism, not independently of it: the criticism determines what is being criticized.

We see here very clearly the relation between a (Platonic) argument and the things it proves. As set up by Socrates the argument (against the thesis that knowledge is perception) lacks an important ingredient; the content of one of its key terms is still undetermined. Yet Socrates argues as if the term had already been defined and comes to a clear and unambiguous conclusion. Thus it was not the argument that produced the conclusion (i.e., the rejection of Theaetetus's thesis that knowledge is perception) but the conclusion (the rejection) produced the argument. As before (see the text to notes 15 and 16 above) an argument—a sequence of sentences moving toward a result—obtains force and even content from a development that occurs outside of it. There are many other examples which conform to this pattern.

A trivial example which I have chosen because of its transparency is the reply to Lactantius's argument against the spherical shape of the earth. The earth, says Lactantius,[18] cannot be spherical because the antipodes would fall down. Here the background is a cylindrical universe. "Up" means a direction parallel to its axis, "down" the opposite direction. Socrates' advice prompts us to replace the cylindrical universe by a central symmetrical one and only now to look for trouble: examining a new idea we first change the world so that it can accommodate the idea. The question if the new world is a possible one comes afterwards. We want to save the spherical shape of the earth. The spherical shape is given— what modifications are needed to retain it in the face of Lactantius' observations? The answer is well-known. We define "up" as "away from the earth," "down" as "toward the center" and get what we want. Rejecting the criticism we redefine its premises.

In the case of Lactantius the spherical shape of the earth was supported by a variety of independent considerations and the adaptive measures therefore only moderately ad hoc. This cannot be said of my second example, which I explored in chapters 6 to 11 of *Against Method* (second and third editions), viz., book 1, chapters 7 through 10 of Copernicus's *De Revolutionibus*. Here Copernicus discusses the dynamical difficulties of the motion of the earth. He introduces the hypothesis that what is part of the earth also moves with it, no matter what other mo-

---

18. "Aut est quisquam tam ineptus qui credat esse homines quorum vestigia sint superiora quam capita. Aut ibi quae apud nos jacet inversa pendere? Fruges et arbores deorsum versus crescere?" *Divinae Institutiones*, 3, "De falsa sapientia." Lactantius was somewhat behind the times, but his argument would have made perfect sense at an earlier age when people, some atomists included, indeed assumed an absolute "up" and "down." See for example Lucretius, *De rerum natura* ii, 205, 227. But cf. i, 1070 ff.

tions it has received. In other words, he introduces a dynamics specially designed to fit the motion of the earth. It is a strange sort of dynamics (for contemporaries averse to occult qualities): it assumes that a distant body can "feel" where the earth is and that it reacts accordingly. Galileo vigorously attacked an analogous principle in the case of the moon (tides). Yet the move was in the spirit of Socrates and the resulting cosmology an advance—according to later views.

Consider next the "paradox" of Einstein, Podolsky, and Rosen. The authors try to show that quantum mechanics permits an "objective" simultaneous determination of the values of noncommuting observables. Using a formidable mathematical apparatus and a "criterion of reality,"[19] they succeed in showing that position and momentum can have simultaneous sharp values. But the "criterion" is just the point at issue. It asserts that the value of a quantity which can be determined "without in any way disturbing [the] system" that contains the quantity, has an "element of reality" corresponding to it, and it is applied when two systems are far enough apart not to influence each other in any way. Now a piece of wood in Australia changes length when length is defined by a rubber unit-meter in Vienna and I go to Vienna, enter the unit-meter institute and stretch the meter. It changes without any physical disturbance—because it is a relation involving other objects. Have quantum-mechanical magnitudes similar properties? Bohr thought so and defined them accordingly. His views may be open to objections (such as an empirical confirmation of Bell's inequality)—but the argument of Einstein, Podolsky, and Rosen is not one of them.

The best modern account of the procedure suggested (but not explicitly described) by Socrates is found in Galileo. Grassi had measured the distance of comets and found them to be beyond the lunar sphere. Galileo pointed out (among other things) that triangulation works only if the nature of the objects triangulated is already settled: it makes no sense to triangulate a rainbow.[20] Ontological or worldview discussion has

19. A. Einstein, B. Podolsky, and N. Rosen, "Can Quantum Mechanical Description of Physical Reality Be Considered Complete?" *Physical Review* 47 (1935): 777–80. The criterion occurs in the first section. Einstein deplored the formalism and ascribed it to Podolsky: A. Fine, *The Shaky Game* (Chicago: University of Chicago Press, 1986), 35 f.

20. Cf. Mario Guiducci, "Discourse on Comets," *The Controversy on the Comets of 1618: Galileo Galilei, Horatio Grassi, Mario Guiducci, Johann Kepler*, trans. S. Drake and C. D. O'Malley (Philadelphia: University of Pennsylvania Press, 1960), esp. 39. Guiducci represents the point of view of Galileo.

to precede the use of counterexamples, it cannot be based on it. But worldview discussion is not different from other kinds of discussion, which means that we can no longer assume discussion-independent and in that sense "objective" arbiters of a debate. This applies even to such apparently trivial cases as "all ravens are black"—the favorite example of naïve falsificationists.

The statement, our logic books explain, is "refuted" by the discovery of a single "objectively" white raven.

Now a raven that has been painted white is white, and even "objectively" and "reproducibly" so—but nobody would regard it as a refuting instance. What we want is "intrinsic" whiteness.

A raven that lost its color in the course of a prolonged sickness is "intrinsically" white—the whiteness came from the inside, not from the outside—but still somewhat problematic. What we want is "normal" color, not exceptions.

Note that the comments made so far have an empirical and a normative component: we assume (empirical component) that there are properties that "belong" to an object and are not "imported"; we also assume (second empirical component) that among them some are "normal," i.e., agree with a criterion that plays an important part in our everyday lives while others do not. We then decide (explicitly, or simply following tradition) to use only ravens which exhibit such properties as counterexamples (this is the normative component). Note also that the statement is not refuted (or confirmed) after these matters have been settled but that settling the matters is part of the process of refutation. This becomes especially clear when we analyze less familiar cases.

Thus consider ravens that became white as a result of evolutionary pressures, or as a result of externally induced genetic changes. The "fundamental dogma" of molecular biology excludes the second case, but how would we deal with it if it occurred? And how shall we deal with the first case? Perhaps by letting color take a backseat compared with criteria and distinctions that are more closely connected with some easily identifiable molecular-biological structures? Again there is an empirical component (close connection) and a normative component (use as counterexamples). At any rate it is now clear (a) that the term "black" in "all ravens are black," though intuitively clear, is ambiguous in the sense that its future use is largely unknown; (b) that it loses some of its ambiguity in the presence of "absurd" counterexamples: as in the case of Achilles a

*The Unfinished Manuscript*

contested view becomes clear only after it has been left behind (clarity, as early anatomists knew, is a property of corpses, not of living things); (c) that what is a counterexample and what not depends on (often unconscious) decisions or rearrangements of thought which are caused by unforeseen developments (*defining* the content of a statement in advance means separating it from the processes which guarantee its continued importance); (d) that the relevant impulses often come from areas outside language (increasing authority of molecular biology and thus decreasing importance of colors as species identifiers); and (e) that for all these reasons "refutation" is a complex process whose result may determine its ingredients rather than the other way around.[21] Again it is not possible to draw a clear and lasting line between the "objective" and the allegedly "subjective" ingredients of the process of knowledge acquisition and of knowledge itself.

This result leads at once to the assertions made toward the end of the first chapter. Thinking and speaking a language, we continuously adapt to the situations we encounter and we change our ideas accordingly. The idea of love we had as children differs from the adolescent idea, which in turn differs from the idea of a great-great-grandmother looking back on a rich rewarding life with a variety of husbands, lovers, children, grandchildren, and dogs. The changes may be abrupt—most of the time they are continuous and hardly noticeable. They are also unforeseen, for nobody can know what events s/he will encounter and how s/he will react to them. Moreover, they grow from the ideas of the moment, which will appear precise and well defined only as long as life is stable and fairly routine: as in the case of anatomy, clarity is a property of corpses.

*3. Reality and Change: An Intermediate Summary*   I CONCLUDE (1) that completely closed cultures (conceptual systems) do not exist; (2) that the openness of cultures is connected with an inherent ambiguity of thought, perception, and action: concepts, for example, are not well-defined entities but much more like forebodings; (3) that the ambiguity can be mobilized by feelings, visions, social pressures, and other nonlinguistic agencies; (4) that these agencies have structure, they can "pressure us to conform with them"

21. For further examples cf. my *Farewell to Reason* (London: Verso, 1987), chap. 6, sec. 2.

(chapter 1 note 18 and text), just as language does and in this way keep linguistic changes meaningful; (5) that argument has power only insofar as it conforms to nonargumentative pressures; (6) that a reality that is accessible to humans is as open and as ambiguous as the surrounding culture and becomes well defined only when the culture fossilizes; also it is only partly determined by research; the basic moves that establish it consist in asserting a certain form of life. I add (7) that the points just made are misleading because they are expressed in terms and dichotomies which suggest a much harder and much more easily manageable subject matter. I shall therefore make them again, this time using a different medium for my arguments.

# INTERLUDE

# ON THE AMBIGUITY

# OF INTERPRETATIONS

*Three Ways of*
*Looking at Traditions* ACHILLES MADE STATEMENTS which sounded strange to his visitors and which also sound strange to some twentieth-century commentators. There is nothing unusual about this situation. People often say strange things. But one can ask them and then the matter will either be clarified, or shelved and forgotten until somebody else starts asking all over again. Outrage, instant dismissal, lack of interest are other possible reactions.

Nobody can ask Achilles, and his visitors did not pursue the matter. They were puzzled, they argued for a while but they soon gave up. This, too, is not unusual. The world is full of garbled messages, unfinished letters, damaged records. Again there are many ways of dealing with the problem, each one having advantages and drawbacks.

The case changes character when it is lifted out of its natural habitat and judged by ideas from a different background. Being confronted with the (occasionally paradoxical) results of such a judgment we, i.e., the distant commentators, can do a variety of things, the following three among them. (1) We accept the judgment; in the special case discussed above we would then agree that Achilles was indeed talking nonsense and we would have to explain how nonsense can anticipate later, and historically identifiable, sense. (2) We change the ideas that lead to the judgment so that Achilles' utterances become meaningful. (3) We draw a distinction between judgments which can be easily incorporated into the practice they comment upon and outside judgments (which seem irrelevant and incomprehensible to those engaged in the practice) and reject the latter.

I shall often return to these three approaches (which I have selected from a much greater variety) and comment on their relative merits. For the moment let me say that they occur in many different fields from the theater to astronomy.

Thus accepting a certain view concerning the nature of factual knowledge or an epistemology, some writers discovered that information produced by their contemporaries did not fit the view and either called it unscientific (Descartes on Galileo), or declared it to be a matter of faith
(Whitehead on Newtonian science). This corresponds to the first approach. Others felt (second approach) that the sciences were essentially
sound but wondered "how scientific knowledge was possible" (Kant).
To obtain an answer they adapted their philosophy to scientific practice
and "rationally reconstructed" the latter. Still others denounced all philosophical interpretations whether critical or supportive and suggested
(third approach) "to see science on its own terms" (Arthur Fine).

Confronted with such a variety most philosophers try to establish one
approach to the exclusion of all others. As far as they are concerned there
can only be one true way—and they want to find it. Thus normative
philosophers argue that knowledge is a result of the application of certain rules, they propose rules which in their opinion constitute knowledge and reject what clashes with them. Pragmatists and the later Wittgenstein, on the other hand, point to the complexity of scientific or, more
generally, epistemic practice and invite us to "look, not to think." The
remaining Kantians, finally, try to get beyond appearances as a machinery that is simple and explains the nature of even the most idiosyncratic
event. Who is right? The case of Achilles shows that this is a rather simpleminded question. Thus Wittgenstein's invitation assumes that events,
which can be identified by inspection, will be missed or misrepresented
by abstract thought. But thought changes looks—which undercuts the
advice. Besides, looking is not a simple matter. The conditions under
which Achilles delivers his report (the tension between his situation and
social requirements; his disappointment) make familiar divisions oper-
ate in unexpected places;[1] they have implications a Wittgensteinian
might ascribe to thought. The remark that Achilles should have looked
without passion reveals a further difficulty. Passion leads to one scenario,
lack of passion to another—that is all we can say when we try to "look,
not to think." The further remark that Achilles did not look but specu-
lated disregards the straightforward character of Achilles' tale. True, he
did not see or hear or smell what he reported, he sensed it in a more "intellectual" way, but in Homer the latter process is as immediate as the

---

1. Cf. chap. 1, text to note 32.

former.[2] Normative rules, on the other hand, may not only fail to find a point of attack in the practice they try to regulate (how do you falsify when there are never any unambiguous falsifying instances?), they may destroy the practice (and perhaps all practices) instead of reforming it. The problem, therefore, is not how to establish a particular approach, the problem is how to use manifest or incipient tendencies to one's own advantage. And even here the choice is not as simple as is suggested by what I have just said. Even an excessively reflective agent is never fully in control. She is already sailing along with one of the tendencies, which means that her choice will appear to her not as a choice but simply as a step on the road to truth. Achilles saw what he saw because he was angry. His anger was not an instrument for exploration which he could apply or drop, according to his inclinations. It was part of his life, therefore part of the tradition to which he belonged, it resonated with a potentially divergent strand of this tradition, recognized it, gave it shape, and, thus, it gave it "reality."[3]

In dealing with Achilles I chose the second approach. I tried to retrace the way in which Achilles supported his assertions, thus making it clear that and why Achilles made sense. And I used "outside" notions such as "language" (in the modern sense), "culture," "worldview," "structure," "ambiguity" to present my findings. *The entire essay, from the examples to the final summing up, is written in this manner.* One must keep this in mind when reading assertions such as the following: cultures contain ingredients which may seem well defined but have much in common with chimeras; they contain open pathways, unknown to anyone; the domains joined by these pathways are often connected like the parts of an Escher landscape; a cultural change that is not the result of plagues, wars, disintegration is started by an impulse, mediated by one of the many conflicting (or Escher-connected) structures the culture contains and comprehended via analogies inherent in the starting point; and so on. All these features (and the stories I chose to make them meaningful) are the results of a particular approach. They are "facts" *as long as the approach fits*

2. Von Fritz, "*Nous, Noein,* and Their Derivatives in Presocratic Philosophy (Excluding Anaxagoras)," *Classical Philology* 40 (1945): 223–42; 41 (1946): 12–34.

3. Similarly, the three approaches I mentioned above may succeed or fail in the sense that they may or may not find resonance in their surroundings. But surroundings change and so do the chances of interpretative systems. It would be a mistake to believe that there is only one way of reading history.

*the group or the tradition that is being addressed.* They dissolve with a different reading of "history," or whatever other entity we use to make sense of events. In a way every reader is in the position of Achilles. Being bombarded by words and harassed by extraordinary events s/he may discover/invent situations undreamt of in popular systems of thought.

The case of Achilles was taken from a text which today is regarded as being largely fictional. This does not lessen its importance. Achilles may never have existed and may never have made the speech that caused so much trouble, then and now. Still, the speech reveals interesting possibilities of conceptual change. Xenophanes, whom I discussed in chapter 2, was a real person. We have extensive quotations from his poems and we are rather well informed about the historical events that surrounded and inspired him. The authors who praise his achievements (see the quotations in chapter 2) employ the first approach. They lift a few passages out of their natural habitat and liken them to modern formulations and ways of reasoning. My counterargument was that transported back into the sixth century B.C. these ways would have been useless as they stand. They have to become part of already existing tendencies and beliefs, which means we have to move from logic to a richer domain of social action.

There is nothing fictitious about Parmenides. He lived, he denied the reality of change and partition (to use modern terms), and he provided arguments (to use another modern term) to support his denial.

Parmenides, too, tells a story. His story is not only well articulated, it is also explicitly subdivided into parts. We can therefore introduce distinctions which otherwise would seem to be artificial and imposed. Aristotle (*Physica* 186a23 ff.) and many analysts after him distinguished a premise, various conclusions, and an argumentative machinery leading from the one to the others. In chapter 3 I adopted their interpretation and criticized Parmenides accordingly. I asserted (a) that Parmenides' denial of change and subdivision is already contained in the premises; (b) that being unsupported by logical reasoning the choice of the premise must be left to a different agency; and (c) that a preference for forms of life where change and subdivision do play a role is one such agency. Following procedure 3 above I would add (d) that Parmenides' arguments are not a continuation of the practice on which they reflect, that they are "outside judgments" and can thus be simply brushed aside. Again it is important to point out that this is not the only possible approach, that there are alternatives, and that they lead to different conclusions.

For example, we may describe Parmenides more in analogy with the cosmological stories that surrounded him. In these stories we have not a premise, but a *beginning* (*chaos* in Hesiod and his Near Eastern predecessors, the *apeiron* in Anaximander, which is both a beginning and a lasting foundation for everything), not an argument but a *pattern of development* (separations and concentrations followed by genealogies in Hesiod, separations and concentrations containing genealogies in Anaximander), not conclusions but stages of development. In Hesiod we have also a revelation: not he but the Muses are the authors of the story. It is here that a different and not purely logical interpretation of Parmenides can start. For Parmenides, too, receives his truth from a Goddess. The proem of his poem explains how, by moving upwards toward the light, he meets the Goddess and hears her report about (logical part) what underlies everything and (physico-historical part) about the way that led up to that "what."[4] Even the logical part allows for a more historical interpretation than the one that is usually given.

There is a beginning. It is sanctioned by a Goddess and it is as devoid of overt content as are *chaos* and the *apeiron*. There is a way of telling the story, and there are results. Given the contemporary use of *einei*[5] the beginning was not difficult to accept—more easily perhaps than *chaos* or the *apeiron* though the first was known also outside philosophy.[6]

The beginning is combined with a story pattern that differs from those of Hesiod and Anaximander but is familiar from the practice of law (see note 34 and text of chapter 2). There are, however, two modifications. First, the legal pattern is radically simplified. Second, it is not applied to criminal cases, but to cosmology. As in Hesiod and Anaximander the pattern produces results one would not have obtained without it. It does not matter that the results sound absurd and paradoxical. The

---

4. For details concerning this particular reading cf. Luciano A. Cordo, *XAOS: Zur Ursprungsvorstellung bei den Griechen* (Idstein: Shulz-Kirchner Verlag, 1989), chap. 5, esp. 351.

5. Cf. Charles H. Kahn, "The Greek Verb 'To Be' and the Concept of Being," *Foundations of Language* 3 (1966): 251 ff., and *The Verb "Be" in Ancient Greek* (Dordrecht: Reidel, 1973).

6. A. Heidel, *The Babylonian Genesis* (Chicago: University of Chicago Press, 1951), 97 ff.

connections and analogies between the story and familiar patterns, assumptions, procedures are strong enough to make them seem important. And indeed—numerous Sophists appropriated the one or the other aspect of the tale, popularized it, and added results of their own.

The existence of analogies of the kind just mentioned (which are cut off by a purely logical account) warns us not to be satisfied with reconstructions of ancient texts that rely on logic and mathematics alone. Such reconstructions may sound "rational"—but they are irrational in the sense that they eliminate features which mobilized the judgment of contemporaries: their "rationality" is for modern consumption only. Given a text we therefore cannot be content with a model that produces one part of it when another part is given and makes sense to a modern reader. The model must consider all elements (for example, it must consider Parmenides' proem) and the textual as well as extratextual implications for the contemporaries.[7]

7. An example will explain what I mean. Examining archaic number concepts, many researchers assume that they share basic elements with later arithmetical notions and that they are numbers because they share these elements. Tables of addition, subtraction, multiplication which are "correct" when interpreted in modern terms seem to confirm the assumption. But such an interpretation cannot explain why counting people was (and still is) supposed to endanger them or why it was (is) assumed that counting a herd can diminish it. Nor can it explain why the oldest known number systems were dyadic and why this feature survived right into the beginning of arithmetic (Pythagoras's theorems about the even and the odd, which apparently led to the discovery of the incommensurability of the square root of two—cf. the proof interpreted in Euclid, *Elements* 10.117). Numbers which have such features and such effects have to be different from the numbers examined by modern number theoreticians. These remarks and the remarks in the text above constitute, as far as I can see, a partial vindication of the observations of Sabetai Unguru, "On the Need to Rewrite the History of Greek Mathematics," *Archive for History of Exact Sciences* 15 (1975): 67 ff. Cf. also van der Waerden's reply in the same journal, vol. 15 (1976): 199 ff.

# CHAPTER FOUR

# BRUNELLESCHI AND THE

# INVENTION OF PERSPECTIVE

*1. The Ugly*
*Madonna of Siena*

FIGURE 2 SHOWS the so-called *Madonna degli Occhi Grossi.* The panel, which is carved in relief out of a single piece of wood, can now be seen in the Museo dell'Opera del Duomo in Siena. In the second half of the thirteenth century it occupied the high altar. It was the most revered object in Siena. It worked miracles. The Sienese believed that it had helped them when, against all odds, they routed the invading Florentine army at the battle of Montaperti (1260).

About 250 years later, Raphael painted his *Madonna del Granduca* (figure 3). Vasari describes the development as follows:

In the first and oldest periods the arts evidently fell a long way short of perfection and although they may have shown some good qualities, were accompanied by so much that was imperfect that they certainly do not deserve a great deal of praise. Then in the second period there was clearly considerable improvement in invention and execution, with more design, better style, and a more careful finish; and as a result artists cleaned away the rust of the old style, along with the stiffness and disproportion characteristic of the ineptitude of the first period [for] it is inherent in the very nature of these arts to progress step by step from modest beginnings and finally to reach the summit of perfection . . . But the most graceful of all was Raphael of Urbino who studied what had been achieved by both the ancient and the modern masters, selected the best qualities from their work and by these means so enhanced the art of painting that it equalled the faultless perfection of the figures painted in the ancient world by Apelles and Zeuxis . . . [His scenes] bring before our eyes sites and buildings, the ways and customs of our own and of foreign peoples, just as Raphael wished to show them . . . his figures expressed perfectly the character of those they represented, the modest or the bold being shown just as they are. The children in his pictures were depicted now with

*The Unfinished Manuscript*

Fig. 2. *Madonna degli
Occhi Grossi,* c. 1250.
Museo dell'Opera
Metropolitana, Siena.
Alinari/Art Resource, NY.

mischief in their eyes, now in playful attitudes. And his draperies are neither
too simple nor too involved but appear wholly realistic. (*Lives of the Artists*
[1550, 2d ed. 1568], trans. George Bull [1965; reprint, Harmondsworth: Penguin, 1979], 84, 252)

Artists, says Vasari, try to represent real things and events. They do not
immediately succeed; held back by ignorance and false traditions, they
produce stiff and crude images of lamentable proportions. But they
gradually improve. Figure 2 (unknown to Vasari) is an early, figure 3 a
later stage in the process. Many Western observers, art historians included, used to agree with this judgment.[1] As late as 1857 the Margaritone of the National Gallery [a picture similar in construction to figure

---

1. Filippo Villani, *De Origine Civitatis Florentiae et eiusdem famosis civibus* (1381/2),
which contains a chapter on painters, told a similar story. The ancient artists, he says, ex-
celled by their skills; after that the arts decayed, skills degenerated, the likeness of nature
was lost or not even strived after. Cimabue summoned back with skill and talent the decayed art of painting—and so on.

Fig. 3.  Raphael,
*Madonna del Granduca*,
c. 1505. Palazzo Pitti,
Florence. Alinari/
Art Resource, NY

2—see figure 4] ". . . was notoriously recommended for purchase to show—as we read in the catalogue—'the barbarous state in which art had sunk even in Italy previously to its revival.'" (E. H. Gombrich, *Ideas and Idols* [Oxford: Phaidon, 1979], 195). Yet, the considerations in this chapter and in the preceding ones suggest a different view.

The imitative view has played an important role in the history of the arts and the sciences. It was accepted by Plato (who criticized painters and poets for imitating the wrong entities), developed by Aristotle (tragedy imitates deep-seated social structures and is therefore "more philosophical" than the most painstaking historical account), taken for granted by Leonardo; it was applied to music by metaphysicians (Kepler) and composers (Monteverdi); it guided Stanislavsky in his search for the right method of training actors; and it found a comfortable home in the sciences (slogan of the unprejudiced scientist who avoids speculation and

Fig. 4.    Margarito of Arezzo. *The Virgin and Child Enthroned, with Scenes of the Nativity and the Lives of the Saints*, c. 1275. © National Gallery, London

"tells it like it is"). It was very influential indeed—but it was not the only motor of scientific or artistic change.

Thus ancient painters, sculptors, art theoreticians suggested and modern technologists still suggest going beyond nature and improving on her. Modern artists, photographers included, reject imitation with particular violence. They may speak of reality—but it is a hidden reality, far removed from common views and impressions. We must also remember that many objects that ended up in museums were meant to decorate, edify, warn, mock, instruct, sell a product, not to imitate, and that the outlines of humans, plants, and animals they might contain functioned as ornaments, as a shorthand, or as magical symbols; they were not maps of reality. Some objects had a rather pedestrian use (jewelry, cassoni, painted trays, clothes), others (such as the Madonna in figure 2, "primitive" images, the statues of late Roman emperors—cf. figure 8) mediated spiritual powers. Imitation itself was occasionally understood in a highly abstract way. Constantine V defined an image as being "identical in essence with that which it portrays." The Eucharist was a true image (of Christ) in this sense; a painted picture was either a mockery or an idol.[2]

Considering this great and almost limitless variety of views and pur-

2. Details in J. Pelikan, *The Christian Tradition*, vol. 2, *The Spirit of Eastern Christianity (600–1700)* (Chicago: University of Chicago Press, 1971), chap. 3, esp. 109.

poses, it would be foolish to subject artistic products to a single criterion and to orient them toward a single aim. A rural crucifix, the scenes that enliven the smooth surface of a piss-pot and an illustration in Grey's anatomy simply do not belong to one and the same line of development.

Another criticism of naïvely imitative philosophies came from historians who took a closer look at periods of alleged decay. For Vasari and his followers such periods are without merit. Early Christian art, for example, is what remains when the classical forms are freed from (to the Christian) objectionable ingredients and then copied by inexperienced and untalented artisans. It is debris, nothing positive. Riegl, who investigated late Roman painting, sculpture, and architecture, found that the debris has distinct and well-defined stylistic features. The features differ from those of classical art. But the determination with which they are imposed and the regularity with which they occur prove them to be the results not of decay, but of positive intentions. They should be judged by these intentions and not by an art-independent "real world."

Riegl's observations are a valuable corrective to a crude progressivism. His positive ideas are another matter. Forms as described by Riegl have much in common with the elements of pure mathematics. Following Riegl (who developed his ideas in the course of a study of the decorative arts) one might say that the arts and pure mathematics coincide in producing patterns but differ in the materials used for displaying the patterns. But pure mathematics is not all of science and the development of forms not all of art: *there are* artists who want to copy nature—and some of them seem to succeed to a surprising degree. How can this additional element be reconciled with a plurality of well-defined and precisely executed artistic styles?

According to Riegl the actions and perceptions of artists are "internally connected" with the block of ideas, institutions, habits that constitute the ideology and with it the worldview of a culture. An artist expresses *visually* what is generally *thought* to be the nature of things; real is what is assumed, thought, and therefore seen to be real at a certain time (*Spätrömische Kunstindustrie* [reprint, Darmstadt: Wissenschaftliche Buchgesellschaft, 1973], 401 f.). This does reconcile imitation and artistic pluralism—but on the basis of an arbitrary and badly founded assumption. I agree that the "objectivity" or worldview-independence (art-independence) of reality cannot be taken for granted—but neither can the close correlation postulated by Riegl. Figure 2 *may* have caught an element of reality that had disappeared by the time of Raphael—but this must be determined by research, not by metaphysical speculations about the "nature of reality." A closer look at the process of imitation shows how complex the situation really is.

*2. A Renaissance* THE BIOGRAPHY of Filippo Brunelleschi, the
*Experiment and* great Florentine architect, contains the follow-
*Its Consequences* ing report of an event said to have occurred in or
about the year 1425:

As for perspective, the first work in which he showed it was a small panel about half a *braccio* [one *braccio* = about twenty-eight inches] square on which he made a picture of the church of San Giovanni in Florence [the Baptisterium]. He painted the outside of the church and as much as can be seen at one glance. It seems that to draw this picture he went some three *braccia* inside the central door of S. Maria del Fiore [the Dome]. The panel was made with much care and delicacy and so precisely, in the colors of the black and white marble, that there is not a miniaturist who could have done better. He pictured in the centre the part of the piazza directly in front of him, and thus, on one side, that which extends towards the Misericordia as far as the arch and the canto de' Pecori; and on the other that from the column commemorating the miracle of St. Zenobius all the way to the canto della Paglia. For the distance, and the part representing the sky, where the boundaries of the painting merge into the air, Filippo placed burnished silver so that the actual air and the sky might be reflected in it, and so the clouds, that one sees reflected in the silver, are moved by the wind when it blows.

The painter of such a picture assumes that it has to be seen from a single point which is fixed in reference to the height and the width of the picture, and that it has to be seen from the right distance. Seen from any other point there would be distortions. Thus, to prevent the spectator from falling into error when choosing his viewpoint, Filippo made a hole in the picture at that point in the view of the Church of S. Giovanni which is directly opposite to the eye of the spectator, who might be standing in the central point of S. Maria del Fiore in order to paint the scene. This hole was small as a lentil on the painted side, and on the back of the panel it opened out into a conical form to the size of a ducat or a little more, like the crown of a woman's straw hat. Filippo had the beholder put his eye against the reverse side where the hole was large, and while he shaded his eye with his one hand, with the other he was told to hold a flat mirror on the far side in such a way that the painting

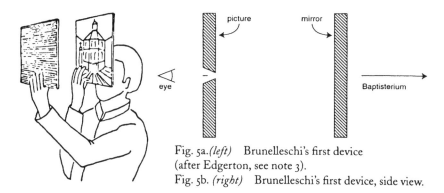

Fig. 5a.*(left)*   Brunelleschi's first device
(after Edgerton, see note 3).
Fig. 5b. *(right)*   Brunelleschi's first device, side view.

was reflected in it. The distance from the mirror to the hand near the eye had to be in a given proportion to the distance between the point where Filippo stood in painting his picture and the church of S. Giovanni. When one looked at it thus, the burnished silver already mentioned, the piazza and the fixing of the point of vision made the scene absolutely real. I have had the painting in my hand and have seen it many times in those days, so I can testify to it.[3]

What we have here is a simple and rather straightforward *scientific experiment*. This, at least, is how the matter can be described today. I enumerate some of the features that characterize the procedure as an experiment.

Brunelleschi *compares* a human product, a picture, with something else. The comparison is not left to the whim of the experimenter, he does not simply take a look, he examines the matter under *rigorously specified conditions* (figures 5a, 5b): he moves to a preassigned place, about three *braccia* inside the entrance of Santa Maria del Fiore, he raises the painting so that it is about five feet from the ground, he looks through a hole in its center, puts a mirror into a well-defined place, removes the mirror and—lo and behold!—there is no change although he now sees the "real" Baptisterium, not a picture of it.

Second, the painting is not the result of trial and error; it is constructed in accordance with definite *rules*. According to Krautheimer[4]

3. The passage is from Antonio di Tuccio Manetti, *The Life of Filippo Brunelleschi* (ca. 1480), quoted in *A Documentary History of Art*, vol. 1 [Princeton: Princeton University Press, 1981], 171 f.; the text offered is not a translation, but an explanatory rendering of the "extremely difficult . . . Tuscan colloquialisms" (167, translator's footnote); I have changed passages using technical language where the text has none). Cf. also S. Y. Edgerton Jr., *The Renaissance Rediscovery of Linear Perspective* (New York: Harper and Row, 1975).

4. Richard Krautheimer and Trude Krautheimer-Hess, *Lorenzo Ghiberti*, vol. 1 (Princeton: Princeton University Press, 1970), chap. 16.

*The Unfinished Manuscript*

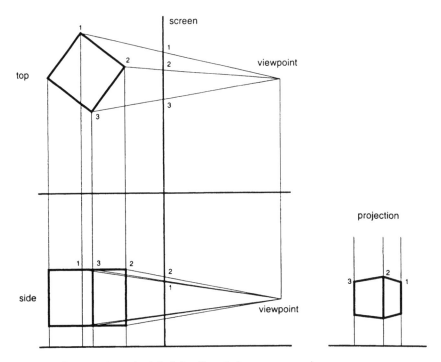

Fig. 6.   Construction principle (after Krautheimer, see note 4).

the rules came from architectural practice and consisted in combining ground plans and elevations drawn to scale with a simple method of projection (figure 6). The rules were soon *systematized* and made part of a new science. In his essay *Della Pittura*, which introduced the science, Leon Battista Alberti defines a picture as a "cross section of the pyramid" formed by the rays extending from the eye to the object.[5] The definition turns picture making into a geometrical problem and the painter into a producer of cross-sections of optical pyramids:

> I say the function of the painter is this: to describe with lines and to tint with colors on whatever panel or wall is given him similar observed planes of anybody so that at a certain distance and in a certain position from the center they appear in relief, seem to have mass and to be lifelike [89].
>
> Our instruction in which all the perfect absolute art of painting is explained will be easily understood by a geometrician, but one who is ignorant in geome-

5. Leon Battista Alberti, *On Painting*, tr. J. R. Spencer (New Haven: Yale University Press, 1966), 52, 48.

try will not understand these and other rules of painting. Therefore I assert the necessity for the painter to learn geometry [90].

Painting rests on *scientific principles,* and progress in painting means either the adoption of such principles where they are not yet in use or an improvement of principles already employed. Alberti also explains how the science of painting is related to other sciences (Alberti's "pyramid" was a key concept of the scientific optics of his time). A painter may proceed intuitively, he may condense the working of his intuition into rules of thumb, but only a study of the laws of geometry can lead to solid achievements. It has also other advantages.

One of the advantages Alberti had in mind was an improvement in the *social standing* of the artists. Traditionally, painting and sculpture were instances of manual labor, on the same level as cooking and the work of blacksmiths. When guilds were founded in Italy in the thirteenth and fourteenth centuries, sculptors and architects were classified with stonemasons and bricklayers, while in England "face painters," coach painters, and house painters all belonged to the Painters-Stainers Company (until the seventeenth century). They were not always fully privileged members of their guild. Major artistic enterprises of thirteenth- to fifteenth-century Italy were planned and supervised in detail by those who had commissioned them, and the artists had to accept their demands. Considering this situation Alberti's treatise, which turned painting into a science, was "just short of subversion."[6] But it achieved its purpose: painters, sculptors, and architects gained in stature. Eventually this led to the foundation of academies and, as is the case with established enterprises, to complaints and objections.

Using modern categories we may say that the objections were of two kinds—scientific and artistic. The "artistic" objections denied that it was the task of artists to simply copy natural objects.[7] The "scientific" objections (which were raised by artists and anticipated corresponding scien-

6. Krautheimer and Krautheimer-Hess, 1:316. For the social position of artists cf. F. Antal, *Florentine Painting and its Social Background* (1947; reprint, Cambridge: Belknap Press of Harvard University Press, 1986), 274 ff., as well as M. Baxandall, *Giotto and the Orators* (Oxford: Clarendon Press, 1971). The story of the rise and development of art academies is told by Nikolaus Pevsner, *Academies of Art, Past and Present* (Cambridge: Cambridge University Press, 1940, new ed., 1973).

7. This view existed already in antiquity. A work of art, it was assumed, improves on the deficiencies of individual natural products and, thus, of nature. Quintilian, *Institutiones oratoriae,* ed. J. J. Murphy [Carbondale: Southern Illinois University Press,

tific developments by more than three hundred years) accepted this task but emphasized what is now called the difference between visual space and physical space. The two spaces coincide for an observer who looks at things in the manner of Brunelleschi, i.e., under special and severely restricted conditions. But an artist, says Leonardo, does not paint for one-eyed observers with their eye in a fixed position and he may not want the surface of his painting to disappear from sight: a picture viewed by people who walk up and down in front of it and recognize it as a picture must be constructed in a different and as yet unknown way.[8]

The development of perspective in the Renaissance and after has much in common with the pattern I outlined in the two preceding chapters. Here as before we have customary ways of ordering and presenting events. The ways are not formalized or tied to rigid rules (except within the confines of particular workshops where secondary painters filled in the outlines drawn by the master and otherwise followed routine), and there is not one order, there are many (example: the difference between the Florentine and the Sienese schools in the early thirteenth century). Now a new schema arrives. It does not force its way in, as in the case of Achilles, it is not immediately nailed down as with Parmenides—it is the result of an almost accidental transference of rules implicit in one practice to another. However, we have ambiguity (soon called "anticipations" of perspective), outside structures, and a state of mind that encourages their use.[9]

---

1987]12.10.7 pointed out that the works of Polyklet gave the human figure "a grace surpassing truth" and criticized Demetrius, who preferred verisimilitude to beauty (12.10.9). Dion of Prusa wrote (*Olympic* 2.167) that "not even a madman would suppose that in his size and beauty the Olympian Zeus of Phidias resembles any mortal being." Even Leonardo, who emphasized the scientific character of painting, observed that the artist can draw "not only the works of nature, but infinitely more than those which nature produces" (quoted in A. Blunt, *Artistic Theory in Italy 1450–1600* (Oxford: Oxford University Press, 1962), 37. Cf. also E. Panofsky, *Idea* (New York: Harper and Row, 1968), for further details and references.

8. *The Notebooks of Leonardo da Vinci*, ed. J. P. Richter, vol. 1 (New York: Dover Publications, 1970), sec. 3. Leonardo knew that the "correct" projection of a sphere is in most cases an ellipse. As far as we know he never depicted a sphere in his paintings. Raphael did depict two spheres in his School of Athens—but he drew their outlines as circles. According to La Guernerie he also made an experiment. In an engraved reproduction of the painting he used the "correct," elliptical shapes—but these were found to be quite unacceptable. Details and analysis in M. H. Pirenne, *Optics, Painting, and Photography* (Cambridge: Cambridge University Press, 1970), 121 f.

9. A vivid account of these elements and their collaboration is given by Giorgio de Santillana, "The Role of Art in the Scientific Renaissance," in *Critical Problems in the*

Having been found, the schema articulates and modifies existing habits, making them more definite and exclusive. Soon a suitable Parmenides comes along who not only describes, but proves, not only shows, but demands. The demands are supported by visual evidence: the schema agrees with "reality," But, as before, the "reality" that provides the proof is not simply given; it is carefully constructed to fit the chosen task. Parmenides' *estin* already lacked what he wanted to eliminate. In the same way Brunelleschi's comparison involved a reduced simile of the object he intended to imitate—the Baptisterium—it did not involve the Baptisterium itself. Later painters, Leonardo and Raphael among them, wanted to work without the reduction (single eye in carefully chosen place) and return to "normal" ways of looking at pictures just as the successors of Parmenides, Democritus and Aristotle among them, wanted to restore the most outstanding features of common sense. Incidentally, such commonsense features had been given profile and thus brought to everybody's attention by the Parmenidean argument itself (there was no separate entity, "common sense," before Parmenides or similarly oriented religious movements had rejected it). Yet, having gone through the experience of perspective, later painters could no longer see their work as part of reality itself (earlier pictures

---

*History of Science,* ed. Marshall Clagett (Madison: University of Wisconsin Press, 1959), 33 ff. Brunelleschi knew the effect of putting familiar objects into unusual situations, large-scale arrangements included. He designed stage sets with complicated machinery, permitted his friend Toscanelli to use the lantern of the Duomo as the apex of a huge sundial, thus turning it into the largest astronomical instrument ever built, and played elaborate tricks on some of his friends. For example, he constructed a new identity for a woodcutter, Manetto di Jacopo Ammanatini, called the Fat One. Playing carefully rehearsed parts and rearranging Manetto's surroundings (apartment, furniture, etc.), Brunelleschi, together with a group of conspirators, all known to Manetto, treated him as if he were somebody else until Manetto started defending the new, illusory identity. Here was proof that things are seen or felt to be "what they are" only in appropriate circumstances and that there exist other circumstances, not at all difficult to arrange, which may dissolve one's sense of self: even the self is not "given," but depends on (unnoticed) projections. It was this insight that enabled Brunelleschi to move from his perspective construction to the corresponding experience and from there to "reality" (details in E. Battisti, *Filippo Brunelleschi* (Stuttgart and Zurich: Belser, 1979), 326 f. Experts in theoretical optics realized only much later (what had been known to their medieval predecessors) that a structure that corresponds to the laws of physical optics does not automatically produce an analogous (realistic, or illusionary) experience. Details in Vasco Ronchi, *Optics: The Science of Vision* (New York: New York University Press, 1957).

of saints were often indications of saintly presence[10]), just as post-Parmenidean thinkers (though not post-Parmenidean common sense) were very much aware of the difference between an object and the perception of it (they also had a great difficulty finding a connection between the two). Still later, abstract concepts (and perspective) were no longer used to imitate, but to construct and thus to introduce entirely new worlds (Masaccio's Trinity and the mannerists are examples). I shall now explore these matters in somewhat greater detail.

*3. Brunelleschi's Painting Interpreted as a Stage*

BRUNELLESCHI EXAMINED HIS PAINTING by checking it against something else. This "something else" was not a building; it was a building as seen with a single eye in a precisely defined place or, as I shall say, it was an *aspect* of a building, an aspect (of an object) being defined as the effect (of the object) on an individual, or a group, or a device (a camera obscura, for example) that approaches, uses, views, analyzes, or "projects" it according to more or less clearly describable, though not always clearly recognized, procedures. Brunelleschi chose an aspect that suited his purpose. His experiment involved two artifacts, not an artifact (the painting) and an art-independent "reality."

Moreover, the chosen aspect (of the Baptisterium) was not compared with the painting itself, but with a carefully arranged mirror image of it. There were two physical objects, the picture here, the building there, with no obvious similarity between them. (Plato, in his diatribe against the arts, used this feature of paintings to good effect.) The building was large, heavy, three-dimensional, made of stone; the picture small, light, its surface two-dimensional, and it was made of wood (a panel) covered by layers of pigment. The objects were projected, the resulting aspects compared and found to be identical. If we want to say that Brunelleschi imitated reality then we have to add that this reality was manufactured, not given. It was "objective" in the sense that, like a statue, its material ingredients existed independently of observations (though not independently of human interference). It was also "subjective," for human experience was an essential part of the arrangement. The best way to describe the situation is by saying that Brunelleschi built *an enormous stage*, containing a preexisting structure (the Baptisterium), a man-made object

10. See e.g. Hans Belting, *Das Bild und sein Publikum im Mittelalter* (Berlin: Gebr. Mann Verlag, 1981).

(the painting), and special arrangements for viewing or projecting both. The reality he tried to represent was produced by the stage set, the process of representation itself was part of the stage action, it did not reach beyond it. Brunelleschi's expertise in the building of stage machinery and in the handling of phenomena such as the phenomenon of personal identity (for details see note 9, above) makes this an adequate description also from his own point of view.

Interpreting artworks as stage sets provides a precise and useful framework for discussing a variety of assumptions about the scope, function, and development of artistic styles.

For example, it refutes the idea that styles arise and change with *necessity* and that the artist, or his employer, has only limited control over the process. Brunelleschi was not swept along by overwhelming historical forces; he prepared every step of his performance. He transferred familiar methods of representation (architectural ground plans and elevations drawn to scale) to a new field, he painted his picture, and he devised an experiment to check its adequacy. Of course, he knew the architectural and stage practices of the time, at least to a certain extent. But he changed both[11] and transferred part of the result to an area that so far had been separated from them. The change might not have been occurred, the transfer (from architecture to painting) might not have occurred (not every architect was as many-sided and idiosyncratic as Brunelleschi), the experiment might have failed (this is a matter of physics and of the physiology of vision, not of historical necessity), its success might have remained an amusing episode without any effect on artistic practice (as did Brunelleschi's transformation of Manetto di Jacopo Ammanatini with its theatrical implications, which appeared only much later, with Artaud, for example)—many accidents had to happen to get systematic central perspective on its way.

Second, the model corrects the idea that it is *mind* and mind alone that imposes a style and that styles are therefore conventions, free from the impediments of the material world. For Brunelleschi used not just his mind and his eyes, he also used a physical contraption and the new aspects appeared only when this contraption had been set up properly. The correction remains in force for traditional art forms whose projective devices (habits of spectators, location, function, and physical

---

11. The construction of the cupola of the Duomo involved a variety of new elements.

arrangement of artworks, institutional, optical, and physiological restrictions of vision) are not consciously introduced and may not even be noticed: what remains unnoticed does not on that account become part of mind just as the unnoticed reactions of the cones do not occur in the mind, but in the fovea.[12]

Conversely, the experiment also corrects the idea that we are dealing here with objective physical laws alone. True, the "picture" can be constructed with the help of rules which are based on (approximations of) the laws of the propagation of light. The camera obscura could be used to demonstrate the action of these laws. But it does not follow that a human, put in the right position, will see things accordingly, not only at the edges whose appearance or disappearance might follow an "objective" pattern[13] but also in the center of the painted area. Ancient architects were familiar with this fact and therefore let their buildings deviate in most interesting ways from the geometric form they wanted to be seen.[14]

Third, some of the contraptions that project a style can be separated, either physically or in thought, from the process of projecting and examined in isolation. For example, we can examine Brunelleschi's equipment and inquire how it affects the perception of things. Or we can examine the social constraints that make people behave in a schematic and lifeless way and study their effect on artistic stereotypes (see text to note 19 below). Dealing with external structures and not with "a specific 'structural form' of the spirit," such an examination is a matter for science, not for an activity that "stands between metaphysical deduction and psychological induction."[15]

12. The physical and physiological aspects of artistic vision are described in Pirenne, *Optics, Painting, and Photography*.

13. As is pointed out by Pirenne, *Optics, Painting, and Photography*, 11, 61.

14. Pirenne, *Optics, Painting, and Photography*, 149 f. Cf. Plato's criticism in *Sophist*.

15. The quotation is from E. Cassirer, *The Philosophy of Symbolic Forms*, vol. 2 (New Haven: Yale University Press, 1955), 11. Cassirer influenced Panofsky and, via him, the discipline of art history. Cf. M. A. Holly, *Panofsky* (Ithaca and New York: Cornell University Press, 1984), chap. 5.

The analysis of a stage into its physical, physiological, psychological, and social ingredients can be stopped at various levels of specificity. Thus in describing Brunelleschi's experiment I omitted the role played by (physical) light, by the eye of the spectator, and by his emotional relation to the things seen. In note 21 I shall discuss a case where the last element becomes important. In the sciences the situation is the same: physicists may use molecules without considering the details of their structure, or they may consider elementary particles without paying attention to the forces that hold molecules together.

Finally, the model of the stage can be readily transferred to the sciences. Like Brunelleschi's setup every scientific experiment involves two series of transformations and a comparison. Nature is transformed to obtain special events, these events are further transformed by data processing devices, scanners, etc. to turn them into evidence which is then compared with the outcome of a transformation of high theory through calculations, computer approximations, phenomenology, etc. An example is the UA1 experiment that started at CERN in summer 1981 and led to the discovery of the W and the Z particles.[16] On the side of the object there was a complicated arrangement leading to proton-antiproton collisions supposed to create short-term appearances of the particles sought together with intricate devices consisting of detectors, computers for multiple tasks (checking the performance of the detectors, selecting possible candidates out of a plethora of events, displaying them for further study by scientists), on the side of the theoreticians there were predictions (from the electroweak theory, specially adapted to the situation of the experiment, often using computers) to guide the selection. Neither "nature" nor its theoretical image were faced directly; they were both transformed by complex and sophisticated processes (cf. figure 7). The notion of a stage containing projective devices omits details but retains the features needed for a general discussion of problems of reality: the role of projections and the associated problem of the projection-independent existence of the aspects projected.

I already mentioned that the model of the stage can be extended to beliefs, theories, and artworks that are not specially manufactured parts of an experiment but are firmly embedded in tradition. In this case the stage is already set. There are established ways of viewing animals, people, mountains, houses—they constitute the reality the artist sets out to explore; and there are equally familiar ways of viewing artworks (knowledge of artistic conventions included)—they determine what people experience when confronted with a statue, or a panel painting, or a fresco. The task of the artist now is this: to create a physical structure which, when approached, or "projected" in the customary manner, produces an aspect similar to one of the familiar aspects of the things represented.[17] I shall

16. Cf. the personal account in Peter Watkins, *Story of the W and Z* (Cambridge: Cambridge University Press, 1986), and the more technical literature suggested there.

17. Note that an aspect so defined includes emotions like fear or respect or, rather, emotions, social tendencies, intentions (such as the intention to obey whatever order is

call projections which are part of a tradition *natural projections,* the aspects they create *natural aspects,* and the structures the artist puts on canvas to produce them *stereotypes.* Again "reality" is part of a stage set, not a set-independent entity, and again the stage set includes nonmental elements. The difference between Brunelleschi and tradition is that while Brunelleschi *controlled* the set, traditional artists *are largely controlled* by it. They rarely notice that a set is involved; like Achilles and his visitors they may believe that they are dealing with "the very objects themselves." Even "progressive" artists think in this way. Not realizing that the social changes that carried them along affected their relations to things, reformers such as Alberti and historians such as Ghiberti and Vasari thought that the ancients had known, that their successors had lost, and that they themselves had rediscovered the unchanging true nature of space, time, matter, and human affairs. They were right in one respect, mistaken in another. The changes of stereotype they deplored did indeed occur. From the third century on, the portraits of saints, bishops, emperors, local administrators, i.e., the statues or pictures (on panels, on coins, in manuscripts) carrying their proper names had indeed lost individual features and had started wearing conventional expressions. "The artists of the early Middle Ages saw, as it were, fewer things in a face than did their predecessors in late antiquity." This was interpreted negatively, as "a sign of graphic impoverishment and decadence."[18] It was assumed that the change was in the artists entirely, not in the world the artists tried to represent and that it led away from (an art-independent) reality. But

> [j]ust as ceremonies and rites replaced spontaneous action at the imperial palace and in official life from the third century on, the physical person of the prince gave way to the august bearer of supreme power, so the image, when it is of the emperor or a dignitary, aimed primarily at a representation of the sovereign, "count," or *silentarius* that is recognizable not so much by his personal traits as by his insignia, his posture, his ritual gesture. And as

given by the appropriate person) and the optical impressions are intertwined in such a way that it makes no sense to try isolating the latter and looking for their "objective" cause. The "aspects" of a panel painting of the Holy Virgin which smiles, or looks threatening, or encourages the petitioner cannot in any way be explained by the laws of geometrical projection.

18. The quotations above and the two quotations that follow are from A. Graber, *Christian Iconography* (Princeton: Princeton University Press, 1968), 65 f.

Fig. 7. Fermi National Accelerator Laboratory, Batavia, Illinois. The Tevatron: The stage set from above, and from the inside. Photos courtesy of Fermilab Visual Media Services.

protocol forbade freedom of movement to the prince and the dignitaries who surrounded him and insisted on perfect immobility of countenance, the art of portraiture, in attempting to represent this way of conceiving the appearance of certain privileged human beings, devised an appropriate plastic formula.

When entering Rome, Constantius II, son of Constantine the Great, stood as immobile as a statue, "looking neither to the right nor left, as if his head were held in a vise" (Ammianus Marcellinus, *Rerum Gestarum Libri xvi*, 16, 10 ff.—see figure 8). Now a rigid face embedded in rigorous ceremony lacks the features that appear in intimate personal contact—but, given the circumstances, the lack is perceived as a positive element, as an expression of power, authority, and permanence. Artists imitated these positive features. Portraits of dignitaries lost individuality because their purpose was to

> convey the idea that the person portrayed is the real basileus, consul, dignitary, or bishop by showing in the portrait that he possesses all the essential characteristics: he has noble and grave features and a majestic mien, he makes the correct gesture, he holds in his hand the insignia or he wears the clothes appropriate to his social situation. We are tempted to say, when we see for instance an image of St. Theodore as a soldier, that it is an image of a Byzantine soldier which resembles other portrayals of Byzantine soldiers. But what we should say is that it is an image of St. Theodore defined iconographically as a Byzantine soldier.

*People themselves* became simpler, their feelings, facial expressions, gestures, and words became more repetitive and rigid just as modern scientists devoted to objectivity and administrative efficiency may lose or consciously suppress, and as a result eventually lose, large parts of what nature provided for them. The change of stereotypes from late antiquity to the early Middle Ages was not restricted to *artworks;* it affected the *objects* of artistic representation as well, brought them closer to art, and thus makes it possible for us to classify the latter (at least in part) as realistic.[19]

19. Commenting on the Byzantine style J. Burkhardt emphasizes its repetitious character, the absence of subjectivity, the grim holiness of Christ, the Virgin, the Saints, and biblical figures. "Motions and attitudes increasingly lose life" (*Cicerone* [Stuttgart: Alfred Kroener Verlag, 1986], 690 f.). However, he adds (693) that though art was never more in fetters than in Byzantine times, people then accepted its principles more readily than ever before or after.

Fig. 8.   Constantius II (337–361),
son of Constantine the Great.
Fourth-century bronze head.

On the other hand it is clear that the stereotypes developed by Brunelleschi and Alberti and praised by Vasari—perspective, natural posture, delicate colors, character, emotions—are not only useless, but unrealistic and downright subversive when the stage has been set in the manner just described. Perspective relates objects to a viewpoint and robs them of their independence (any bum can reduce the emperor to the size of an ant by assuming a suitable position), "delicate colors" do not inhere in the objects but are created by suitable illumination, "natural postures" are accidental, and, for many purposes irrelevant cross-sections of behavior that would disrupt the orderly and socially real stage action. *Both* the artists who tried to represent power and authority *and* the individuals who attempted to embody them were forced to develop stereotypes (of posture, arrangement, behavior) incompatible with the principles of an optical realism.

The situation is very clear in Egyptian art. The falcon on the palette of King Narmer (first dynasty—figure 9) is "animated" and lifelike, the falcon of King Wadj (also first dynasty—figure 10) stern and stylized—

Fig. 9. Slate palette of King Narmer. Egyptian Museum, Cairo. Giraudon/Art Resource, NY.

but it makes no sense to speak of decay. The execution is excellent, the "stiffness" (a term applied by Vasari to the "old style," which, according to him, preceded Cimabue and Giotto) or, as we should rather say, abstractness of the design is not proof of ineptitude, but of concentration. Besides, the "lifelike" style did not disappear—it continued to be used for everyday scenes (figure 11). Both styles issued from the same workshops and were practiced by the same artists: the workshop of Thutmosis in Tell el Amarna (the ancient Achet-Aton) has realistic masks of live models (figure 12) side by side with more schematic representations (figure 13). During the reign of Amenophis IV (1364–1347), who replaced the old priestly religion with a cult of the sun and the rigid forms of traditional art with an exuberant expressionism (which mercilessly exposed his unregal face and his weak and emaciated body) the stereotypes changed twice. The first change, just described, occurred about four years after his ascent to power (figure 14a), the second change restored the status quo (figure 14b). Artistic competence played no role in the

Fig. 10.   Funeral stone of King Wadj. Louvre, Paris. Giraudon/Art Resource, NY.

process; institutional, religious, political, and personal transformations did. We can here see, almost as in a laboratory, how aspects depend on the social set and the associated forms of projection.[20]

Thus, the function of the arts was not always to "give a true representation of reality"—the arts quite often had the purpose to aid the viewer by magic, and/or to strengthen social cohesion (later on I shall try to show that the same is also true of certain "scientific" views).

The conflict between geometric-chromatic congruence and more "spiritual" values is clearly described in a Greek text known as the Apocryphal Acts of Saint John, which has been attributed to the second century A.D. and to Asia Minor. The text tells how Lycomedes, a pupil of

20.  Details are found in H. Schaefer, *Principles of Egyptian Art* (Oxford: Clarendon Press, 1974). Plato, who was interested in permanence, praised the Egyptians for "canonizing" stereotypes (*Laws* 656d) and criticized perspective for "getting hold of only a small part of the objects, and that a phantom" (*Republic* 598b).

Fig. 11.    Servant crushing grain. Egyptian Museum, Cairo. Giraudon/Art Resource, NY

John, secretly invited a painter to John's house and asked him to paint John's portrait. John discovered the painting but, having never seen his own face, failed to recognize it and thought it to be an idol. Lycomedes brought a mirror and John, comparing mirror and painting said:

> As the Lord Jesus Christ lives, this portrait is like me, yet, my child, not like me but only like my fleshly image. For if the painter who has here imitated my face wants to draw it in a portrait, he will be at a loss [needing more than] the colors which you now see on the boards . . . and the position of my shape and old age and youth and all the things that are seen with the eye. But you, Lycomedes, should become a good painter for me. You have the colors which he gives you through me, he, who himself paints all of us, even Jesus, who knows the shapes and the appearances and the postures and types of our souls. But what you have done here is childish and imperfect: you have drawn a dead likeness of the dead. (A. Grabar, *Christian Iconography* [Princeton: Princeton University Press, 1968], 66 f.)

In other words: an optical realism that concentrates on colors and geometry leaves out life and the soul. Or, to speak in a less naïvely realistic

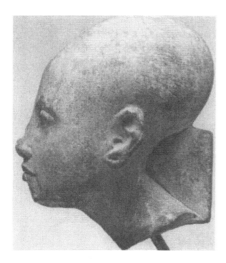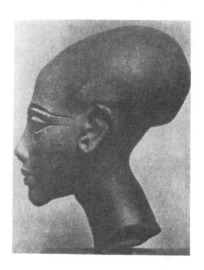

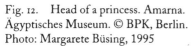

Fig. 12.   Head of a princess. Amarna.
Ägyptisches Museum. © BPK, Berlin.
Photo: Margarete Büsing, 1995

Fig. 13.   Head of a functionary.
Amarna. Ägyptisches Museum,
Berlin. Giraudon/Art Resource, NY

manner: stereotypes that concentrate on the visual appearance of the surfaces that define a human face cannot accommodate the stereotypes of spirituality.[21]

To sum up: artistic imitation (and artistic production in general) occurs in a sometimes well-defined, but often very loose context, it takes place on a "stage." The stage contains the artwork, the methods of imitation, projective devices for creating the aspects to be imitated, as well as these aspects themselves. Whatever "reality" is being taken into consideration is manufactured by the stage, it changes with the setting, and

21. The objection that the optical image contains everything and therefore also contains "soul" overlooks the fact that, like all images, it is the result of special projections which may have omitted relevant cues. The existence of such cues was shown by an examination of blind people who had regained their sight. The first things they recognized were not, as gestalt psychologists had assumed, simple structures (spheres, squares), but objects to which they had an emotional relation. "A girl who was an animal lover identified her beloved dog first of all." A. Ehrenzweig, *The Hidden Order of Art* (Berkeley and Los Angeles: University of California Press, 1967), 13. Leonardo (*"Paragone,"* or, *First Part of the Book on Painting*, quoted in E. G. Holt, ed., *A Documentary History of Art*, vol. 1, *The Middle Ages and the Renaissance* (Princeton: Princeton University Press, 1981), 277, emphasizes that "painting extends only to the surface of bodies; perspective deals with the increase and decrease of bodies and their coloring, because an object as it recedes from the eye loses in size and color in proportion to the increase of distance. Therefore painting is philosophy, because philosophy deals with the increase and decrease through motion as set forth in the above proposition."

Fig. 14a.  *(left)* Akhenaton
(Amenophis IV), c. 1360 B.C.
Äegyptisches Museum, Staatliche
Museen, Berlin. Art Resource, NY.

Fig. 14b.  *(below)* Tutankhamen and his
queen (from a throne), c. 1350 B.C.
Egyptian Museum, Cairo. Giraudon/Art
Resource, NY.

Fig. 15. This figure may at first look like a meaningless array of fragments but is transformed when recognized.

artistic stereotypes change with it. The elements of the stage are physical bodies, institutions, customs, powerful beliefs, economic relations, physical processes such as light and sound, physiological processes such as color vision, the mechanisms creating the perception of sound and musical harmony and many other events.

Stages are either newly built, or they are part of a tradition. In the first case the aspects and projections may have to be learned just as a traveler must learn the language and the customs of an alien country. For example, people had to learn (and people unfamiliar with optical illusionism still have to learn—cf. text to figure 15) to see perspective both in the world and in pictures based on perspective constructions. An artificial stage can be absorbed by history. Brunelleschi's construction and Alberti's systematization gave rise to a new artistic tradition, new ways of relating to pictures and things, and corresponding new perceptions. Photography, movies, and rock videos caused further massive changes.

Stage sets that are part of history have the same structure as artificial stage sets. The "reality" traditional artists try to imitate is constituted by devices of the same general nature as those introduced by Brunelleschi. Having become a habit, the devices are not noticed as separate constitutive entities. Both the artists and their public seem to face reality directly, and without mediation. It is this (unavoidable and very powerful) impression of immediacy and easy access that underlies naïve realism (cf. my comments on the "inside view" made in connection with Achilles' complaint). The impression dissolves once alternative ways of creating order gain the upper hand. They make manifest what has been hidden before, activate its inherent ambiguity, and use it to effect change: comprehensive stages that were built into customs and beliefs and were therefore removed from awareness become explicit frameworks within other stages which now lack definition. The history of perspective contains many examples of this development.

Again I have to point out that in speaking of "stages," "projections," "aspects" I made things far more definite than they are. The terminology seems appropriate when applied to Brunelleschi's procedure, for here we have indeed something that is best described as the "setting up of a stage." It imposes rather than reveals a pattern when extended to traditions whose development is largely unplanned. It is quite correct to observe that these traditions may have had their own ideas of the function of art and that even where imitation reigned supreme the aim may not have been to imitate the surfaces of relaxed individuals but to show their social position. Given certain turning points, the observation may even be exact. But we go too far when inferring a "system" and, after that, a general relativity of artistic efforts. For the exactness we may on occasion encounter is part of a process that overcomes it and replaces it with an entirely different arrangement. It was not there before the process started, it does not survive its termination. This means, of course, that the real situation that existed when the process started was open, indefinite, and capable of modification. Trying to catch it by a "system" and then inferring a general relativism would be as sensible as trying to define the shape of a body of water by the shape it assumes when frozen and inferring a radical difference between water, ice, and steam.

Only a few of the artists who try to discredit the impression of immediacy are opposed to immediacy itself. What they object to is the immediacy of a style at variance with their own. The aspects of the latter (which are familiar and immediate to them though unfamiliar to others) are interpreted as part of a real world that exists independently of their effort.[22] This is perfectly in agreement with what I said about the case of

22. Alberti took it for granted that the space he described and whose two-dimensional construction he explained was true space and that attitudes implying a different space had no foundation in reality: "And here I cannot help inquiring what would be the reason of a very whimsical, though very old persuasion, which is firmly rooted in the minds of the vulgar, that a picture of God or of some saint in one place shall hear the prayers of the votaries, when in another place, the statue of the very same God or saint shall be utterly deaf to them. Nay, and what is still more nonsensical, if you do but remove the very same statue, for which the people used to have the highest veneration, to some other station, they seem to look upon it as bankrupt, and will neither trust it with their prayers, or take the least notice of it. Such statues should therefore have seats that

Achilles: the "inside view" indeed confronted him with a new and as yet unrealized reality. Scientific realists do the same. Starting on their journey of exploration they "project." Finding coherence in their projections they combine them into a world. Disregarding the projecting mechanisms which by now have become second nature, they assert the objective existence of this world. This is naïve realism all over again—only tied to special and relatively unfamiliar stage sets. How can such a procedure deny the reality of the forces emanating from figure 2?

*4. Two Versions of Realism: Naïve and Relativistic*

THE QUESTION (at the end of the previous section) raises two kinds of problems, one factual, the other dealing with concepts.

The factual problems concern the role of artworks (theories, sets of beliefs) and the reactions of the contemporaries. Not all artworks and not all parts of science were built for imitation. The models of mediaeval astronomy, though extremely useful, made no claim to objective truth, while churches, statues, mosaics, frescoes, panels combined functions which today are assigned to separate domains. For example, the objects that adorned the house of God not only set the stage for the central event of Holy Mass—the daily transformation of the eucharistic wafer into the body of Christ and of the sacred wine into His blood—they also reminded the faithful of their obligations and they taught the illiterate the elements of the Roman creed. Being approached in a personal way, they might give advice, affect individual decisions, and cause miracles. During processions, mobile representations of Christ, the Virgin, the saints often assumed the role of the characters themselves.[23]

How were the objects treated by the public? Were they treated as they are today, as the admirable productions of outstanding individuals who are permitted to do anything, as long as there are suitable critics around to praise their products, and if so, were they praised, criticized, or simply disregarded? Were they praised for their richness (material, expensive colors—properties very important for those buyers who wanted to show off), for the surprising tricks they contained (extreme foreshortenings,

are fixed, eminent, and peculiar to themselves." *De re aedificatoria*, bk. 7, chap. 17, quoted in Joan Gadol, *Leon Battista Alberti* (Chicago: University of Chicago Press, 1973), 150 f.

23. Details in Bruce Cole, *The Renaissance Artist at Work* (London: John Murray, 1983), and, for a more special area, H. Belting, *Das Bild und sein Publikum im Mittelalter* (Berlin: Gebr. Mann Verlag, 1981).

for example), the opportunities for learned comment they offered to the cognoscenti? Were they praised for their lifelikeness, their ability to deceive the beholder, or were they revered as sources of miraculous power, revealing a reality the viewers believed in but could not reach by themselves? All these elements played a role, the relative weight of the roles changed through the ages, and it is difficult to disentangle them in a particular case.

Above all—how were the objects perceived? Meeting a stranger and recognizing him as a friend one has not seen for a long time drastically changes his appearance—from an unfamiliar ancient relic to a not so unfamiliar and certainly not so ancient person. A similar switch occurs when one mistakes one's mirror image for a real person, different from oneself, and then suddenly realizes one's error; an unpleasant countenance may change into a very attractive one. Writers (Pirandello, for example) gave vivid descriptions of transitory phenomena of this kind while artists (Kokoschka) stabilized some of their stages, thus creating the impression of a hidden reality which, though often concealed, has a powerful effect upon our lives. For many viewers figure 15 at first sight looks like a meaningless jumble of shapes: unusual methods of representation have unusual results. Pictures we ourselves can read with ease are not recognized by people unfamiliar with their use.[24] Now it is known that the "uneducated" contemporaries of Giotto were confused by his novel style, which omitted details and imposed a coherent spatial frame. They preferred pictorial narrations without too much abstraction and

24. "Take a picture in black and white," writes Dr. Laws, a Scottish missionary who was active in Nyasaland, "and the natives cannot see it. You may tell the natives: 'This is a picture of an ox and a dog'; the people will look at it and look at you and that look says that they consider you a liar. Perhaps you say again, 'Yes, this is a picture of an ox and a dog.' Well, perhaps they will tell you what they think this time! If there are boys around, you say: 'This is really a picture of an ox and a dog. Look at the horn of the ox [drawing of lines, as in the text above], and there is his tail!' And the boy will say: 'Oh yes, it is a dog.'" Quoted in J. B. Deregowski, "Illusion and Culture," in *Illusion in Nature and Art*, ed. R. L. Gregory and E. H. Gombrich (New York: Scribner, 1980), 163. The article contains further examples of this kind. Figure 15 at first sight has the same effect on many Western observers; it looks like a meaningless jumble of shapes. Some instruction is needed, and a definite shape will arise. The appearance of the shape is not independent of the instruction. Nor does it make sense to say that it already was there and just waited to be perceived. It was produced by the instruction but, after production, remained an independent entity to which descriptions and secondary pictures had to be adapted.

without the tyranny of a unified space.[25] Who can say how strong faith and the physical surroundings of a thirteenth-century Tuscan peasant influenced his ways of seeing pictures and things? Figure 2 may well have been perceived as representing or even possessing properties invisible in a beautiful but bland and almost kitschy painting such as figure 3. This history of the panel strongly supports such a view.

Now assume that our historical curiosity has indeed discovered the widespread existence of judgments of this kind—does it follow that the properties perceived are real properties (and not only erroneous perceptions) of the world we inhabit, or that they once were such properties but then disappeared? And how are they related to the properties perceived and represented by later generations? This question introduces the second, the conceptual (or "philosophical") group of problems. To deal with it I shall add some further examples to those already introduced.

MANY PEOPLE WOULD AGREE that figure 16, while not exactly a work of genius, succeeds in capturing the features of Michael Faraday, the physicist. Using the picture we might be able to recognize him on the street or to identify him in a police lineup. Figure 17 gives no such guidance. It is colorful, emotive, bound to please individuals with a special taste, but there seems to be no situation, whether real or imagined, it can be said to represent. It also seems to be much too irregular to interest mathematicians concerned with pure form. It confirms the once popular idea, shared by some intellectuals, that modern art is empty expression, without any relation to things like mountains, dogs, or people.

Now turn to figure 18. It is an illustration from Descartes' essay *L'homme* and explains how (part of) human vision works. The essay itself

25. Petrarca (*Testament*, ed. T. F. Mommsen [Ithaca: Cornell University Press, 1957], 78 ff.) and Boccaccio (*Decamerone* 6.5), repeating a phrase used by Pliny about Zeuxis, said that Giotto's art astounded the connoisseur but left the ignorant cold. Later on, when the plague and financial disasters had reduced the power of the upper bourgeoisie and increased the influence of "uneducated upstarts" (G. Duby, *L'Europe au Moyen Age* [Paris: Flammarion, 1984]), the art of Giotto was felt to be too modern and too abstract. The Dominicans, who were well acquainted with popular tendencies, then started encouraging the older and more posterlike idioms. Details in Antal, *Florentine Painting*, 159 ff. The attitude of the humanists is described in Baxandall, *Giotto and the Orators*. Baxandall's book makes it clear that the humanists' reports, being tied to special preoccupations (projections), have no "objective" validity. Of Brunelleschi's contemporaries, only Masaccio fully grasped and mastered perspective.

*The Unfinished Manuscript*

Fig. 16.  Michael Faraday (1791–1867); drawing by George Richmond.

continues an old tradition intent on finding a stable world behind perceptual change. According to the tradition (which, in the West, goes back to the Pythagoreans and especially to Plato),[26] reality resides in quantitative structures and laws. Figure 18 represents part of the reality Descartes wants to uncover. Lacking this information an opponent of figure 17 might have extended his objections to figure 18. Possessing it he will classify 18 with 16, and not with 17. What is the reason for his change of mind?

The reason is that both 16 and 18 are now seen to correspond to situations that are independent of their production. This does not mean that lines (straight lines like those drawn in 18 or curved lines like some of the features of 16) exist in the world even if nobody ever draws them or thinks of them; it means that the drawing of lines produces, or makes visible, features which have a life of their own and thus are suitable objects for comparison or "imitation." Is figure 17 a pattern without an object or

26. Cf. his critique of the arts in *Republic*, bk. 10.

Fig. 17.  Fernand Léger
(1881–1955), *Woman
in Blue* (1912). Oeffentliche
Kunstsammlung Basel,
Kunstmuseum. Donation
by Dr. h.c. Raoul La
Roche, 1952. Photo:
Oeffentliche Kunstsamm-
lung Basel, Martin Bühler

can it, too, be said to "correspond" (in the manner just explained) to certain features of our world?

Figure 19 is the result of a repeated application of a simple algorithm.[27]
It is a possible object for imitation by structures such as figure 17. Figure
17 thus *might* be a true image of a deeper reality that clashes with common sense but need not be rejected on that account. A mere *look,* or a
look accompanied by familiar judgments cannot decide the question—
we need additional information. More especially, we need information
about the structure and the projecting mechanisms of the stage surrounding figures 16, 17, and 18.

27. From B. B. Mandelbrot, *The Fractal Geometry of Nature* (Oxford: Freeman, 1982),
plate C5 and explanation. Mandelbrot's book develops mathematical structures for
processes which at first sight seem arbitrary, chaotic, and without "objective relevance."

Fig. 18.    Illustration from *L'homme* by René Descartes, first published (posthumously) in 1664.

The required information usually contains two parts: results and ideology. In the case of figure 2 the results are the miracles and the attitude on the spectators; the "ideology," a worldview illuminated by faith. Scientific examples have the same structure. Thus behaviorism can point to successes in situations that seemed inaccessible to other procedures (these are the results) and it prides itself on giving an "objective" account of human affairs (this is the ideology).

Giving an "objective" account means that people are approached, or "projected" in a special way. Like Brunelleschi, behaviorists set up a stage, containing subjects, researchers, experimental equipment, and possible descriptions; they project their subjects, call the projected aspects facts, and identify them with a stage-independent reality. But the very presence of a stage and of methods of projection throws doubt on the identification. After all, a different stage (tradition) such as common sense with its reliance on empathy and introspection produces different aspects leading to different ways of explaining, and reacting to, human behavior. Why should one type of aspect (figure 16) be regarded as "real" while another (figure 17) receives no such dignity?[28]

28.  A classical discussion of a special version of this dualism is P. E. Meehl, *Clinical vs. Statistical Prediction* (Minneapolis: University of Minnesota Press, 1954). Note that any reference to predictive success assumes that the things predicted within one system of projection are more important than the achievements of the other: results decide a question only in conjunction with the value judgments concerning their importance.

Fig. 19.    Fractal, created by Mark R. Laff and V. Alan Norton. Courtesy of International Business Machines Corporation (see note 27).

Moreover, in figure 16, the representation-part of the commonsense stage itself appears only in special circumstances, as we have seen (text before note 25). Again the question arises why the aspects of one particular stage (such as the stage set up by a special scientific investigation) should be regarded as real while those of another should count as appearances only (and remember that "aspect" as defined in the section on Brunelleschi includes physical processes of the most varied kind).

LET ME REPEAT THE CONTEXT of the question. I am not yet asking which of the many things we know are real and which are not. I assume that the world is being approached, or "projected," in a special way, that its representations (stories, diagrams, pictures, perceptions, theories) receive an analogous treatment, and that aspects arise in this manner. The notion of an aspect is ontologically neutral—it simply means the result of a procedure without any implications as to its (degree of) reality. I add that projections may become a habit, may even be built into our constitution and thus remain unnoticed. For example, we "project" when looking at the world in a wide-awake state, with our senses in good order, and in "normal" lighting conditions—but we are not aware of this fact. Special aspects such as perspective, or the images seen in a microscope, which initially create difficulties can be learned and stabilized. All this is a triviality for evolutionary epistemologists, neurophysiologists, linguists, artists, even for some physicists (complementarity).[29] Having stated my assumption I point out that aspects which emerge from different stages occasionally clash and thus cannot be *simultaneous* parts of one and the same stage-independent reality. It is still possible to say, and many realists, both in the arts and in the sciences, do say, that the aspects that emerge from *some* stages are "real" while the aspects of others are not. For example, scientifically inclined realists will say that stars conceived as complicated material systems with a long history are real while Gods, though important ingredients of historically identifiable states, are not. They are not "out there"— they are nothing but products of our projecting mechanisms. And asked for their rationale they give the two kinds of reasons already mentioned: results and ideology. Do these reasons decide the matter?

They decide the matter for people who value the results and accept the ideology. But now the problem returns. Every tradition that survived major difficulties and affects large groups of people has "results" which are important to its members and a worldview (ideology) that unites the details, explains and "justifies" them. Realism as just described cannot reduce this variety except in an arbitrary dogmatic and, let us admit it, rather naïve way. Relativism takes it at its face value. Which view shall we adopt?

*5. Their Common Background*   THE QUESTION ASSUMES THAT relativism and realism are clear alternatives; one of them is correct, the other is not. But relativism and realism share an important assumption: the traditions (stages, means of projection) which relativists regard as equally truthful messengers of reality and

29. A good survey is P. Watzlawick, *Wie wirklich ist die Wirklichkeit?* (Munich: Piper, 1976). Cf. also his reader *Die erfundene Wirklichkeit* (Munich: Piper, 1981).

which realists devalue to enthrone their favorite stereotypes are conceived as being well defined and clearly separated. They are different worlds (or sham worlds, for the realist), they develop according to their own inner dynamics, judge matters according to their own well-defined standards, and do not get entangled with each other. If this assumption fails, then both (naïve) realism and relativism cease to be acceptable.

There are traditions which agree with the assumption. Examples are isolated societies that have adapted to a stable (cultural or ecological) niche or groups, communities, cultures endowed with an exclusive philosophy and held together by rigid institutions. The aspects created by such stage sets have indeed the tendency to coagulate into well-defined and stable stereotypes. But most groups, societies, traditions not only interact, they are built for interaction. Moreover, their systems of projection are ambiguous in the sense that small (intellectual, physical, emotional) pressures can transform them in a fundamental way: not being well-defined is part of their nature. An example will explain the situation.

The criteria which identify a natural language do not exclude change. English does not cease to be English when new words are introduced or old words given a new sense. Every philologist, anthropologist, sociologist who presents an archaic (primitive, exotic, etc.) worldview, every popular science writer who wants to explain unusual scientific ideas in ordinary English, every surrealist, dadaist, teller of fairy tales or ghost stories, every science fiction novelist, and every translator of the poetry of different ages and nations knows how first to construct, out of English *words* an English-*sounding* model of the pattern of usage he needs and then to adopt the pattern and to "speak" it. A trivial example is Evans-Pritchard's explanation of the Azande word *mbisimo* designating the ability of their poison oracle to see faraway things. In his book *Witchcraft, Oracles, and Magic among the Azande*[30] Evans-Pritchard "translates" *mbisimo* as 'soul'. He adds that it is not soul in our sense, implying life and consciousness, but a collection of public or "objective" events. The addition modifies the use of the word "soul" and makes it more suitable for expressing what the Azande had in mind. Why "soul" and not another word? "Because the notion this word expresses in our own culture

30. *Witchcraft, Oracles, and Magic among the Azande,* abr. ed. (Oxford: Clarendon Press, 1975), 55.

is nearer to the Azande notion of *mbisimo* of persons than any other English word"—i.e., because of an *analogy* between the English soul and the Azande *mbisimo*. The analogy smoothes the transition from the original to the new sense; we feel that despite the change of meaning we are still speaking the same language. Now if a conceptual change like the one just described does not go through a metalanguage but stays in the language itself (in which case we would speak of changing the properties of things rather than the use of words), and if it is not only a single term but an entire conceptual system that is being received, then we are faced with a transformation which, though starting from a well-defined mode of projection and proceeding through a series of steps which, though surprising, conform to the mode, prepares an entirely different stage.

Chapter 1 describes such a transformation. As mentioned there, archaic Greek thought contained analogies for what Achilles was trying to express. Divine knowledge and human knowledge, divine power and human power, human intention and human speech (an example used by Achilles himself: 312 f.) were opposed to each other as Achilles opposed personal honor and its collective manifestations. Guided by the analogies, Achilles' audience could be drawn into his way of seeing the tension and could discover, as Achilles did, a new side of honor and of morality. The new side was not as well defined as its archaic predecessor—it was more a foreboding than a concept—but the foreboding engendered new linguistic habits and, eventually, a new linguistic stage with (relatively) clear new concepts (the frozen concepts of Parmenides and Zeno are endpoints of this line of development).

Forebodings are frowned upon by philosophers who want speech to be clear throughout, by scientists who insist on formal precision and full empirical backing, and by artists who want objects, styles, and modes of representation to be transparent and well defined. The example shows that such demands can block conceptual (artistic) change and cannot explain it when it occurs. The transition from one style to another or from one linguistic stage to another remains a mystery. Taking the isolated and frozen traditional concepts as a measure of sense, we are even forced to say that Achilles speaks nonsense.[31] But measures of sense (or of imi-

---

31. Cf. A. Parry's account in "The Language of Achilles," *Transactions and Proceedings of the American Philosophical Association* 87 (1956), and my own comments in *Against Method*, rev. ed., 214.

tative adequacy) are not rigid and unambiguous and their changes not so unfamiliar as to prevent the listeners from grasping what Achilles has in mind. Speaking a language or explaining a situation, after all, means both *following* rules and *changing* them; it is a whole whose subdivisions (logic and rhetoric; logical and emotive changes) are not inherent in the process itself and whose concepts contain elements logicians are liable to exclude from the domain of thought.

There are authors who agree with this but who still use the platonizing terminology of the majority of logicians to make their point. Needless to say, they have to express themselves in a rather paradoxical manner. For example, they are forced to assert that speaking a language goes through situations where language loses its informative character and consists in making noises only. Given the terminology this is of course a perfectly adequate way of describing what they have discovered. Moreover, their results can be supported by instances of the most varied kind. Little children learn a language by attending to noises which, being repeated in suitable surroundings, gradually assume meaning. Commenting on the explanations which his father gave him about questions of logic, Mill wrote in his autobiography: "The explanations did not make the matter clear to me at the time; but they were not therefore useless; they remained as a nucleus for my observations and reflections to crystallise upon; the import of his general remarks being interpreted to me, by the particular instances which came under my notice *afterwards*."[32]

32. *Essential Works of John Stuart Mill*, ed. Max Lerner (New York: Bantam Books, 1965), 21. For the whole problem, cf. also Heinrich von Kleist's illuminating essay "Über die allmähliche Verfertigung der Gedanken beim Reden," in *Meisterwerke deutscher Literaturkritik*, ed. H. Mayer (Stuttgart: Henry Goverts Verlag, 1962), 743: "I believe that many great speakers didn't know at all what they were going to say when they started talking. But they received courage from the conviction that the circumstances and the resulting excitement of the emotions would lead to the needed abundance of ideas, and so they trusted their luck when uttering their first words. I remind you of the 'thunderbolt' Mirabeau set against the master of ceremonies who had returned after the dissolution of the last meeting with the king on June 23 (in which the latter had ordered the estates to disperse) . . . and had asked them if they had heard the king's order? 'Yes,' Mirabeau answered, 'we have heard the king's order'—I am absolutely sure that starting in this humane way he did not yet think of the bayonets with which he concluded. 'Yes, dear Sir,' he repeated, 'we have heard it'—one notices that he does not yet know what he wants. 'But what gives you the right,' he continues—and now, suddenly, an immense idea starts growing in him—'what gives you the right to relay orders to us? We are the representatives of the nation.' That was what he needed! 'The nation gives orders, it does not re-

Saint Augustine advised parsons to teach the formulae of the faith by rote, adding that their sense would emerge as a result of prolonged use within a rich, eventful, and pious life. Theoretical physicists and "pure" mathematicians love playing around with uncomprehended formulae until a lucky combination makes everything fall into place (in the case of the quantum theory we are still waiting for this lucky combination). Achilles produced habits of speech and thought which eventually gave rise to new and more abstract conceptions of honor, virtue, being.[33] Note also that the analogies that establish a connection with the status quo need not be restricted to concepts in the strictly logical sense but can include images, associations, feelings, moods which enter the terms involved and modify their semantic core. Concepts such as justice, or beauty, even the concept of number are constantly being changed in this way. Absorbing the perceptions and the moods of a new era they first become ambiguous and then flip over into new meanings. This is how processes which for a logician are mere noise, mood, or perception can affect even the most advanced stages of speaking a language.[34]

One scientist who was aware of the complex nature of explanatory talk and who used its elements with superb skill was Galileo. Like Achilles, Galileo gave new meanings to old and familiar words; like Achilles he presented his results as parts of a framework that was shared and understood by all (I am now speaking of his change of basic kinematic and dynamical notions); unlike Achilles he knew what he was doing and he tried to conceal the lacunae that remained and the nonsemantic elements he needed to carry out the change. He succeeded beyond expectation; by creating the impression that his moves occurred on a well-defined stage

---

ceive them'—reaching at once the summit of impudence. 'And to be completely clear'—and only now he finds expression for the resistance his soul is prepared to offer: 'tell your king that we shall not leave our places except when driven away by bayonets'—and then, satisfied, he sat down." Only Kierkegaard shows similar understanding for the conditions of "reasonable action" and the role of the ideas, emotions, images that accompany it.

33. A more balanced account that is not tied to a sharp distinction between semantic and psychological (sociological) elements will of course point out that we are not dealing with a complete absence of meaning but with a situation in which clear and well-defined elements (feelings, ideas, perceptions) are replaced by vague but still not utterly meaningless forebodings.

34. For further details on the transition from Homer's aggregate universe to the substance universe of the pre-Socratics cf. chap. 16 of my *Against Method*, rev. ed. (London: Verso, 1988).

with stable projecting mechanisms and well-defined concepts, he deceived everybody, and perhaps even himself.[35]

WHAT DO EXAMPLES SUCH AS THESE teach us about "reality" and how do they affect the assumption, formulated at the beginning of the present section, that traditions, their projecting mechanisms, and their stereotypes are well-defined and clearly separated?

Using modern terms (which have no equivalent in Homer!) we may say that archaic honor was not merely a "subjective" or "mental" event, but an "objective" relation between an individual and the society to which he belonged. The "personal" honor Achilles has in mind seems to be even less a matter of opinion for it can exist undiscovered, and independently of human judgment. This it can indeed—but only for people for whom such a mode of existence makes sense. There must be linguistic means for describing judgment-independent events and these means must have a point of attack in the world, i.e., in our case, in human beings. For example, the events we now call elements of consciousness must congeal into wholes that provide starting points for self-examination (a collection of itches is not yet a mind). The "archaic mind" was not structured in this way—strictly speaking it did not even exist.[36] A realist who praises Achilles for his "discovery" is therefore in the same predicament as is Achilles himself: he ascribes meaning to meaningless noise and substance to scattered events and leaves the process of knowledge acquisition and knowledge change as it is understood by logicians and epistemologists. His remarks are as empty as those of Achilles—unless the actual development leads to a stage that favors the assertions of both. But when that has happened we have not only new views, we have also a new world (minds where minds did not exist before), which means that a diagnosis of epistemic progress (which assumes that our *ideas* have moved closer to a stable *reality*) loses its point.

Relativism (archaic honor is real for archaic people, personal honor for stages where a personal I comes to the fore) is limited for an analogous reason: the stages which relativists regard as equally valid projectors of truth and reality contain ambiguities which, when becoming manifest,

---

35. For details cf. *Against Method,* chaps. 6 and 7.
36. Cf. ibid., chap. 16, as well as my *Farewell to Reason* (London: Verso, 1987, chap. 4, sec. 4.

dissolve all relativistic judgments. I conclude that relativism and realism, while perhaps leading to approximate accounts of special stages of a complex development, omit important features of these stages and fail when applied to the development itself.

It is not at all out of place to compare social matters such as those just described with the features of a rather interesting abstract physical theory, viz. quantum mechanics. Quantum mechanics (in some of its formulations) admits situations in which familiar properties behave in a familiar way. They correspond to historical stages with more or less well-defined features.[37] Here relativism and realism make sense, though only approximately. But quantum mechanics also describes situations which exclude any of the familiar "classical" properties (Schrödinger's cat is a popular example). These situations correspond to the intermediate stages of historical development, when moods, feelings, mere noises seem to replace clear and well-defined thought. The subdivision is of course artificial—we are never completely clear about what we say and the noises we utter are never completely without meaning. The reason is easy to see; being in the world we not only imitate and constitute events, we also reconstitute them while imitating them and thus change what are supposed to be stable objects of our attention. This complex interaction between what is and the (individual and social) activities leading up to what is said to be makes it impossible to separate "reality" and our opinions in the way demanded by realists. The conclusion is obvious where social objects and events (such as the existence, or nonexistence, of minds) are concerned. It also applies to physical events as I shall try to show in the next section.[38]

37. Examples: the "aggregate" stage and the "substance" stage as described in chapter 1, and in *Against Method*, rev. ed., chap. 16.

38. Here ends the manuscript. There were no further sections in the manuscript papers. *Ed.*

# PART TWO

---

# ESSAYS ON THE MANUSCRIPT'S THEMES

# {1}  REALISM AND THE
# HISTORICITY OF KNOWLEDGE

T HE PROBLEM DESCRIBED in the title is not new. In the West it arose with the Pre-Socratics; it was formulated by Plato and Aristotle, dismantled by the rise of modern science, and reappeared with quantum mechanics and the increasing strength of historical (as opposed to theoretical) accounts of knowledge. Stated briefly, the problem consists in the following question: How can information that is the result of idiosyncratic historical changes be about history-independent facts and laws? To examine the problem, I would like to explore two assumptions and the difficulties that arise from their common use.

*1. The Assumptions*      THE FIRST ASSUMPTION is that the theories, facts, and procedures that constitute the (scientific) knowledge of a particular time are the results of specific and highly idiosyncratic historical developments.

Many events support this assumption. The Greeks had the mathematics and the intelligence to develop the theoretical views that arose in the sixteenth and seventeenth centuries, yet they failed to do so. "Chinese Civilization," writes J. Needham, "had been much more effective than the European in finding out about Nature and using natural knowledge for the benefit of mankind for fourteen centuries or so before the scientific revolution,"[1] and yet this revolution occurred in "backward"

Published in *Journal of Philosophy* 86, no. 8 (August 1989): 393–406. Reprinted with kind permission. A less developed version of this article entitled "Gods and Atoms: Comments on the Problem of Reality" was published in Dante Cicchetti and William M. Grove, eds., *Thinking Clearly about Psychology*, vol. 1 (Minneapolis: University of Minnesota Press, 1991). One page in section 2 of the present article duplicates material in part I, chapter 2, section 4.

1. N. J. T. M. Needham, *Science in Traditional China* (Cambridge: Cambridge University Press, 1981), 3, 22 ff. For details, cf. the sections of Needham et al., *Science and*

Europe: deficient, not good, knowledge led to better knowledge. Babylonian astronomy concentrated on special events, such as the first visibility of the moon after a new moon, and constructed algorithms for predicting them. Trajectories, a celestial sphere, and the considerations of spherical trigonometry played no role in these algorithms. Greek astronomy posited first physical (Anaximander), then geometrical trajectories (Eudoxos, Apollonius), and built on them. Both methods were empirically adequate[2] and capable of refinement (equants, excenters, epicycles in Greek astronomy, polygons instead of step functions or zigzag functions in the Babylonian alternative). Cultural factors, not empirical adequacy, determined the survival of the one and the disappearance of the other.[3]

Recent studies added impressive evidence to these general observations. Historians of science examining the microstructure of scientific research, especially of modern high-energy physics, found many points of contact between the establishment of a scientific result and the conclusion of a complicated political treaty. Even experimental "facts" turned out to depend on compromises between different groups with different experiences, different philosophies, different financial backing, and different bits of high theory to support their position. Numerous anecdotes confirm the historico-political nature of scientific practice.[4]

*Civilisation in China* (Cambridge: Cambridge University Press, 1956 ff.), esp. vol. 5, pt. 7, *The Gunpowder Epic.* A brief but suggestive comparison of simultaneous developments in Greece and China is found in articles by Jacques Gernet and Jean-Pierre Vernant in Vernant, *Myth and Society in Ancient Greece* (New York: Norton, 1988), 79 ff.

2. Interesting comparisons are found in O. Neugebauer, *A History of Ancient Mathematical Astronomy* (New York: Springer Verlag, 1975), 56 (anomaly of the solar motion), 301 ff. (spherical astronomy), 308 f. (lunar motion), and 341 f. (use of zigzag functions for the description of periodic phenomena). Cf. also B. L. van der Waerden, *Science Awakening*, vol. 2, *The Beginnings of Astronomy* (Groningen: P. Noordhoff, 1954), chap. 7.

3. The idea that Greek astronomy was inherently superior reflects part of the evidence which sets stable theories on the one side (Babylon) and a rapidly growing research program on the other (Greece). But the growth here and the stability there did not come from empirical successes or obstacles; it came from different social conditions; cf. G. E. R. Lloyd, *The Revolution of Wisdom* (Berkeley and Los Angeles: University of California Press, 1987), chap. 2.

4. Details in Andrew Pickering, *Constructing Quarks* (Chicago: University of Chicago Press, 1984); Peter Galison, *How Experiments End* (Chicago: University of Chicago Press, 1987); and the September 1988 issue of *Isis.* Anecdotes are unpopular among logicians who see science propelled by the rational use of rational principles. They become important once scientific practice has been reintegrated with history.

The second assumption is that what *has been found* in this idiosyncratic and culture-dependent way (and is therefore formulated and explained in idiosyncratic, ad hoc, and culture-dependent terms) *exists* independently of the circumstances of its discovery. In other words, we can cut the way from the result without losing the result. I shall call this assumption the *separability assumption.*

The separability assumption, too, can be supported by a variety of reasons. Indeed, who would deny that there were atoms long before the scintillation screen and mass spectroscopy, that they obeyed the laws of quantum theory long before those laws were written down, and that they will continue to do so when the last human being has disappeared from Earth? And is it not true that the discovery of America, which was the result of political machinations set in motion by false beliefs and erroneous estimates and which was misread by the great Columbus himself, did in no way affect the properties of the continent discovered? It does not matter that some philosophers have objected to sweeping judgments such as these and have urged us to let science speak for itself, for "science itself" is full of sweeping judgments such as those just mentioned.[5]

So far we have seen the two assumptions that underlie the problem; now for the difficulties.

## 2. The Difficulties

THE SEPARABILITY ASSUMPTION is not only part of science but also of nonscientific traditions. According to Herodotus and Greek common sense (sixth and fifth centuries B.C.), Homer and Hesiod did not create the Gods, they merely

5. Arthur Fine, in an interesting and highly informative book, *The Shaky Game* (Chicago: University of Chicago Press, 1986), invites us "to let science stand on its own and to view it without the support of philosophical 'isms'" (9): to "try to take science on its own terms, and try not to read things into science" (149). Having proposed a similar view in *Against Method* (London: New Left, 1975), I have much sympathy for Fine's position, but I cannot accept it as final. Science is not the only enterprise producing existence claims, and scientific arguments not the only "complex network of judgments" to "ground" such claims (153). Are we to become complacent relativists and accept as existing whatever people lay on us in a sufficiently complicated way ("*complex* network of judgments"), or should we not rather choose among networks and find reasons for our choice? At any rate, there are enough "objective" existence claims contained in molecular biology, the theory of evolution, cosmology, even in high-energy physics (cf. Pickering, *Constructing Quarks*, 404, on the "retrospective realism" of scientists) to bring the difficulties of the next section right into the center of science.

enumerated them and described their properties. The Gods had existed before and they were supposed to live on, independently of human wishes and mistakes. The Greeks, too, thought they could cut the way from the result without losing the result. Does it follow that our world contains particles and fields side by side with demons and Gods?

It does not, the defenders of science reply, because Gods do not fit into a scientific worldview. But if the entities postulated by a scientific worldview can be assumed to exist independently of it, then why not anthropomorphic Gods? True, few people now believe in such Gods and those who do only rarely offer acceptable reasons; but the assumption was that existence and belief are different things and that a new Dark Age for science would not obliterate atoms. Why should the Homeric Gods—whose Dark Age is now—be treated differently?

They must be treated differently, the champions of a scientific worldview reply, because the belief in Gods did not just disappear; it was removed by argument. Entities postulated by such beliefs cannot be said to exist separately. They are illusions, or "projections"; they have no significance outside the projecting mechanism.

But the Greek Gods were not "removed by argument." The opponents of popular beliefs about the Gods never offered reasons that, using commonly held assumptions, showed the inadequacy of the beliefs. What we do have is a gradual social change leading to new concepts and new stories built from them. To see this, consider two early objections against the Gods of Homer. The first objection occurs in a well-known lampoon by Xenophanes,[6] the traveling philosopher. Ethiopians, he says, picture their Gods as snub-nosed and black; Thracians as blue-eyed and blond. And he adds:

> But if cattle, or lions, or horses had hands, just like humans;
> if they could paint with their hands, and draw and thus create pictures
> then the horses in drawing their Gods would draw horses; and cattle
> would give us pictures and statues of cattle; and therefore
> each would picture the Gods to resemble their own constitution
> 
> (*Die Fragmente der Vorsokratiker,* Diels-Kranz, eds.
> [Zurich: Weidmann, 1985], fragment B15)

6. *Die Fragmente der Vorsokratiker,* Diels-Kranz, eds. (Zurich: Weidmann, 1985), fragment B16.

In contrast, the "true" God is described as follows:

> One God alone is the greatest, the greatest of Gods and of men
> not resembling the mortals, neither in shape nor in insight.
> Always without any movement he remains in a single location
> as it would be unseemly to walk now to this, now to that place.
> Totally vision; totally knowledge; totally hearing.
> But without effort, by insight alone, he moves all that is.

(ibid., 23–26)

Can these lines be expected to convert a convinced "Ethiopian" or "Thracian"? Not a chance! For the obvious reply is: "You, Xenophanes, seem to dislike our Gods—but you have not shown that they don't exist. What you did show is that they are tribal Gods, that they look and act like us and that they differ from your own idea of God as a superintellectual. But why should such a monster be a measure of existence?"

Xenophanes' mockery can even be inverted, as we see from a remark by Timon of Phleios, a pupil of Pyrrho:

> Xenophanes, semi-pretentious, made mincemeat of Homer's deceptions,
> fashioned a God, far from human, equal in all his relations,
> lacking in pain and in motion and better at thinking than thought.[7]

The upshot is that Xenophanes' mockery works only if the entity he wants to introduce already impressed itself on the minds of his contemporaries; *it formalizes a historical process, it cannot bring it about.* (I suspect that almost all "arguments" that "advance thought" have this feature.)

A second early "proof" against polytheism which was recorded by the Aristotelian school[8] makes the situation even clearer. According to the proof,

> God is either one or many.
> If many, then either equal or unequal.
> If equal, then like the members of a democracy.
> But Gods are not members of a democracy, hence unequal.
> But if unequal, the lesser is not a God.
> Hence,
> God is one.

7. Sextus Empiricus, *Outlines of Pyrrhonism*, I, 224.
8. *On Melissus, Xenophanes, Gorgias* 977a14 ff.

The proof assumes that being divine means the same as having supreme power. This was not true of the Homeric Gods. Again, the proof hits its aim only after the needed change in the notion of divinity has occurred: *history, not argument, undermined the Gods.* But history cannot undermine anything, at least not according to the separability assumption. This assumption still forces us to admit the existence of the Homeric Gods.[9]

It does not force us, scientific realists object, because a belief in anthropomorphic Gods, though perhaps not removed by reason, never was a reasonable belief. Only entities postulated by reasonable beliefs can be separated from their history. I shall call this the *modified separability assumption.*

Now, making reasonability a measure of separable existence means assuming that things are adapted to criteria of existence and not the other way around. Scientific practice does not conform to this assumption, and rightly so. Birds are said to exist because we can see them, catch them, and hold them in our hand. The procedure is useless in the case of alpha particles, and the criteria used for identifying alpha particles do not help us with distant galaxies or with the neutrino. Quarks, for a time, were a doubtful matter, partly because the experimental evidence was controversial, partly because new criteria were needed for entities allegedly incapable of existing in isolation ("confinement").[10] We can measure tem-

---

9. Marcello Pera objected that I used too narrow a notion of "argument" to make my point. I reply that by "argument" I mean any story that can be told in a relatively short time, has the purpose of showing that the Homeric Gods do not exist, and does this in an "intellectual" way, i.e., using propositions and not procedures such as terror, hypnosis, etc.

It should also be added that the ancient Gods were not creatures of fantasy, but a living presence. In hardheaded Rome, they even participated in the political process. Cf. *Against Method,* rev. ed. (London: Verso, 1988), chap. 16 (Homeric Gods); Robin Lane Fox, *Pagans and Christians* (New York: Norton, 1987), pt. 1, chap. 4 (title: "Seeing the Gods") (late Roman Empire); and Donald Strong, *Roman Art* (London: Penguin, 1982) (Roman Republic and Empire through the fourth century). The God of Judaism, the God of Christianity, the God of Islam, and the saints are even better examples of the ways in which divine Beings affected human life: they profoundly influenced the arts, the philosophy, and the politics of the West and this not only in retrospect, but for those who worked out the details.

10. Andrew Pickering, *Constructing Quarks* (Chicago: University of Chicago Press, 1984), chap. 4. The experiments used the Millikan arrangement. G. Holton, "Subelectrons, Presuppositions, and the Millikan-Ehrenhaft Dispute," in *Historical Studies in the Physical Sciences,* vol. 9, ed. R. McCormich, L. Pyenson, and R. S. Turner (Baltimore:

perature with a thermometer; but this does not get us very far. The temperature in the center of the sun cannot be measured with any known instrument, and the temperature of events such as the first few seconds of the universe was not even defined before the arrival of the second law of thermodynamics. In all these cases, criteria were adapted to things and they changed and proliferated when new things entered the stage. The criteria of acceptability of beliefs changed with time, situation, and the nature of the beliefs. To say that the Homeric Gods do not exist because they cannot be found by experiment or because the effects of their actions cannot be reproduced is, therefore, as foolish as the remark, made by some nineteenth-century physicists and chemists, that atoms do not exist because they cannot be seen. For if Aphrodite exists, and if she has the properties and idiosyncrasies ascribed to her, then she certainly will not sit still for something as silly and demeaning as a test of reproducible effects (shy birds, people who are easily bored, and undercover agents behave in a similar way).

Let me dwell a little on this point. "A wise man," says Hume (and many scientists agree), "proportions his belief to the evidence."[11] Aristotle elaborates:

> Of all beings naturally composed some [the stars] are ungenerated and imperishable for the whole of eternity, but others are subject to coming-to-be and perishing. It has come about that in relation to the former . . . the studies we can make are less, because both the starting points of the inquiry and the things we long to know about present extremely few appearances to observation. We are better equipped to acquire knowledge about the perishable plants and animals because they grow beside us. Both studies have their attractions. Though we grasp only a little of the former, yet because the information is valuable we gain more pleasure than from everything around us, just as a small and random glimpse of those we love pleases us more than seeing many other things large and in detail. But the latter, because the information about them is better and more plentiful, *take the advantage in knowledge. . . .*[12]

Johns Hopkins University Press, 1978], 161 ff.), shows how Millikan and Ehrenhaft, using different versions of this arrangement, and evaluating their data in different ways, got different results for the charge of the electron. In practice an experimental setup is a heap of equipment wrapped in an often very complex ideological blanket. Pickering (*Isis* 72 [1981]) describes the blanket in the case of quarks.

11. *An Enquiry Concerning Human Understanding,* sec. 10, *On Miracles.*
12. Aristotle, *De partibus animalium* 1.5.

But the epistemic power ascribed to areas of research does not conform to this principle. Artisans at all times possessed detailed information about the properties of materials and of their behavior under the most varied of circumstances, whereas theories of matter from Democritus to Dalton were considerably less specific and their connections with the evidence much more tenuous.[13] Yet questions of reality and of suitable methods of discovery were often formulated in their terms, not in artisan terms. The information of artisans did not even count as knowledge. More recently, hydrodynamics and the theory of elasticity, though connected with experience in many ways, received a place far below abstract mechanics (Lagrange, Hamilton). Even comprehensive sciences, such as chemistry and biology, for a long time were assigned a secondary role in the hierarchy of the sciences. When the time scale of mid-nineteenth-century geology and biology exceeded the estimates of the solar age as determined by physicists (such as Helmholtz and Kelvin) and corresponding estimates of the cooling of the earth's surface, prevalence was given to the highly conjectural numbers of the latter.[14] We have to conclude that the authority of a subject, like its shape, is as much a result of idiosyncratic historical developments. True, the persistence of the atomists paid off: modern elementary particle physics, quantum chemistry, and molecular biology would not exist without it. But these achievements could not be foreseen, and an appeal to unknown and as yet inconceivable future effects can also be applied to the Gods. Thus, the modified separability assumption cannot help us to get rid of Gods either: having decided to separate history and existence, we must separate the existence of the Gods even from the most "advanced" scientific argument.

13. Cyril Stanley Smith, *A Search for Structure* (Cambridge: MIT Press, 1981), distinguishes between theories of matter such as the atomic theory and a knowledge of materials. He describes how the latter arose millennia before the former, was more detailed, and was frequently impeded by theoretical considerations. In an exhibition (items presented and explained in *From Art to Science* [Cambridge: MIT Press, 1980], he showed the enormous amount of information contained in the products of early artisans. Norma Emerton, *The Scientific Reinterpretation of Form* (Ithaca: Cornell University Press, 1984), describes the battle between form theories (which were fairly close to the practice of the crafts) and atomism (which was not) and comments on the methods the atomists used to remain on top.

14. J. D. Burchfield, *Lord Kelvin and the Age of the Earth* (New York: Hill and Wang, 1975).

Result: neither the separability assumption nor the modified separability assumption can make us accept atoms but deny Gods. A realism that separates being and history and yet assumes that being can be grasped by history is forced to populate being with all the creatures that have been considered and are still being considered by scientists, prophets, and others. To avoid this abundance, some philosophers and scientists made the following move: Scientific entities (and, for that matter, all entities), they said, are projections and thus tied to the theory, the ideology, and the culture that postulates and projects them. The assertion that some things are independent of research, or history, belongs to special projecting mechanisms that "objectivize" their ontology; it makes no sense outside the historical stage that contains the mechanisms.[15] Abundance occurs in history, it does not occur in the world.

But not all projections are successful. The "scientific entities" mentioned above are not simply dreams; they are inventions that went through long periods of adaptation, correction, and modification, and then allowed scientists to produce previously unknown effects. Similarly, the Gods of antiquity and the God of Judaism, the God of Christianity, and the God of Islam who replaced them were not just poetic visions. They also had effects. They influenced the lives of individuals, groups, and entire nations. Gods and atoms may have started as "projections"— but they received a response, which means they apparently succeeded in bridging the gulf that naïve realists had erected between being and their own historical existence. Why has this response disappeared in the case of (the) Gods? Why is it so powerful in the case of matter?

*3. The Power of Science* WHAT I HAVE SAID SO FAR is simple and straightforward. Even a well-constructed argument cannot remove the overwhelming impression, however, that anthropomorphic Gods were killed by the arrival, first, of philosophical and, then, of scientific rationalism. I shall therefore approach the matter again, and from a slightly different angle. Gods—and this now means the Homeric Gods as well as the omnipotent creator-God of Judaism, Christianity, or Islam—are not only moral, but also physical powers. They cause thunderstorms, earthquakes, and floods; they break the laws

---

15. Fine has shown that this was also the way in which Einstein understood his own "realism" (*The Shaky Game,* chap. 6).

of nature to produce miracles; they raise the sea and stop the sun in its course. But such events are now either denied or accounted for by physical causes, and remaining lacunae are being swiftly closed by research. Thus projecting the theoretical entities of science, we remove Gods from their position of power and, as the more fundamental scientific entities obey time-independent laws, show that they never existed. Many religious people have accepted this argument and diluted their creed until it agreed with this philosophy.

Now, the fact that science dominates certain areas of knowledge does not by itself eliminate alternative ideas. Neurophysiology provides detailed models for mental processes; yet the mind-body problem is being kept alive, both by scientists and scientifically inclined philosophers. Some scientists even demand that we "put[ ] mind and consciousness in the driver's seat,"[16] i.e., that we return to them the power they had before the rise of a materialistic psychology. These scientists oppose the elimination and/or reduction of prescientific psychological ideas and entities. There is no reason why Gods whose numinous aspects always resisted reduction should be treated differently.

Second, a reference to basic time-independent laws works only if the modern accounts of divinely caused events such as thunderstorms, earthquakes, volcanic eruptions, etc., can be reduced to them. But there exist no acceptable reductions of the required kind. Special fields introduce special models whose derivability from basic physics is assumed, not shown. Already Descartes was aware of this situation. Commenting on the richness of the world, he professed himself unable to reduce the properties of special processes such as light to his own basic principles and used a variety of "hypotheses" instead.[17] Newton's[18] discussion of the properties of motion in resisting media abandons the deductive style of his planetary astronomy. It is "almost entirely original and much of it is false. New hypotheses start up at every block; concealed assumptions are employed freely and the stated assumptions sometimes are not used at all."[19] Modern researchers in the area explicitly reject reductive demands: "It is as ridiculous to deride continuum physics because it is not obtained from nuclear physics as it would be to reproach it with lack of founda-

16. R. Sperry, *Science and Moral Priority* (Westport, Conn.: Greenwood, 1985), 32.
17. *Discourse on Method*, pt. 6.
18. *Principia*, bk. 2.
19. C. Truesdell, *Essays in the History of Mechanics* (New York: Springer, 1968), 91.

tion in the Bible."[20] The general theory of relativity for a long time was connected with the known planetary laws by conjecture, not by derivation (calculating the path of Mercury, one added the Schwarzschild solution to the results of prerelativistic perturbation theory without having shown from basic principles that both adequately described the situation in the planetary system). The connection between quantum mechanics and the classical level is quite obscure and has only recently been dealt with in a more satisfactory manner.[21] Meteorology, geology, psychology, large parts of biology, and social studies are even further removed from unification. Thus, instead of a multitude of particulars firmly tied to a set of time-invariant basic laws, we have a variety of approaches with unifying principles looming indistinctly in the background—a situation quite similar to what happened in Greece after Zeus's victory over the Titans.[22]

Third, we are far from possessing a single consistent set of fundamental laws. Basic physics, the alleged root of all reductions, is still divided into at least two principal domains: the world of the very large, tamed by Einstein's general relativity, and the quantum world which itself is not yet completely united. "Nature likes to be compartmentalized," wrote Dyson describing this situation.[23] "Subjective" elements such as feelings and sensations, which form a further "compartment," are excluded from the natural sciences, though they play a role in their acquisition and control. This means that the (unsolved) mind-body problem affects the very foundation of scientific research. Science has large lacunae; its alleged unity and comprehensiveness are not a fact but a (metaphysical) assumption, and those of its projections which work come from isolated areas and thus lack the destructive power ascribed to

20. *Encyclopedia of Physics*, vol. III/3 (New York: Springer, 1965), 2.

21. For a survey of problems and a sketch of possible solutions, cf. Hans Primas, *Chemistry, Quantum Mechanics, and Reductionism* (New York: Springer, 1981). The book also contains a detailed discussion of the relation between chemistry and basic physics.

22. Hesiod, *Theogony* 820 ff.

23. *Disturbing the Universe* (New York: Harper and Row, 1979), 63. Superstring theoreticians try to overcome the remaining multiplicity, but all they have so far succeeded in doing is devising a language in which they can talk about everything without ever arriving at any concrete results. In the words of Richard Feynman: "I do feel strongly that this is nonsense" (in P. C. W. Davies and J. Brown, *Superstrings* [New York: Cambridge University Press, 1988], 194. Besides, even if superstring theoreticians succeeded in uniting basic physics, they would still have to deal with chemistry, biology, consciousness, etc.

them. They show how certain sections of the world respond to crude approaches; they give us no clue about the structure of the world as a whole.

Finally, as the most fundamental and most highly confirmed theory of present-day physics, the quantum theory rejects unconditional projections and makes existence depend on special historically determined circumstances. Molecules, for example, the basic entities of chemistry and molecular biology, do not simply *exist*—period—they *appear* only under well-defined and rather complex conditions.

If one still insists that the bits and pieces of science that are flying around today are superior by far to the analogous collections of a past age—a live nature, whimsical Gods, etc.—then I must refer back to what I said earlier: the superiority is the result of having followed a path of least resistance. Gods cannot be captured by experiment, matter can. This point, incidentally, plays a role also within the sciences. "The great success of Cartesian method and the Cartesian view of nature," write R. Levins and R. C. Lewontin commenting on the significance of the recent advances in molecular biology,

> is in part a result of a historical path of least resistance. Those problems that yield to the attack are pursued most vigorously, precisely because the method works there. Other problems and other phenomena are left behind, walled off from understanding by the commitment to Cartesianism. The harder problems are not tackled, if for no other reason than that brilliant scientific careers are not built on persistent failure. So the problems of understanding embryonic and psychic development and the structure and function of the central nervous system remain in much the same unsatisfactory state they were fifty years ago, while molecular biologists go from triumph to triumph in describing and manipulating genes.[24]

E. Chargaff writes as follows:

> The insufficiency of all biological experimentation, when confronted with the vastness of life, is often considered to be redeemed by recourse to a firm methodology. But definite procedures presuppose highly limited objects.[25]

The insufficiency of science vis-à-vis the Gods could not be expressed more clearly.

24. Levins and Lewontin, *The Dialectical Biologist* (Cambridge: MIT Press, 1985), 2 f.

25. Chargaff, *Heraclitean Fire* (New York: Rockefeller University Press, 1978), 170.

*4. Dogmatism,*
*Instrumentalism,*
*Relativism*

THERE EXIST VARIOUS WAYS of dealing with this situation. One is to disregard it and to continue describing the world in accordance with one's own pet metaphysics. This is the attitude of most scientists and scientific philosophers. It is a sensible attitude; it was the attitude of the educated Greeks and Romans who retained their Gods amidst a flurry of philosophical objections. It does not solve our problem.

Instrumentalists react by dropping the second assumption. They do not drop it absolutely ("nothing exists"), however, but only with respect to certain entities.[26] A confrontation with alternative ontologies revives the problem.

Relativists accept the first assumption but relativize the second: atoms exist *given* the conceptual framework that projects them. The trouble here is that traditions not only have no well-defined boundaries, but contain ambiguities and methods of change which enable their members to think and act as if no boundaries existed: potentially every tradition is all traditions. Relativizing existence to a single "conceptual system" that is then closed off from the rest and presented in unambiguous detail mutilates real traditions and creates a chimera.[27] Paradoxically this is done by people who pride themselves of their tolerance toward all ways of life.

Relativists are right, however, when asserting that the temptation to project certain entities (Gods, atoms) increases in some circumstances, and is diminished in others. Given the auspicious circumstances, the entities do indeed "appear" in a clear and decisive way. More recent developments in the interpretation of quantum mechanics suggest regarding such appearances as phenomena (Bohr's term) that transcend the dichotomy subjective/objective (which underlies the second assumption). They are "subjective," for they could not exist without the idiosyncratic conceptual and perceptual guidance of some point of view (which need not be available in explicit form); but they are also "objective": not all ways of thinking have results and not all perceptions are trustworthy. New terminology is needed to adapt our problem to this situation.

26. This was realized by Duhem, who described a certain stage of the debate between realism and instrumentalism in astronomy as a battle "between two realist positions." *To Save the Phenomena* (Chicago: University of Chicago Press, 1969), 106.

27. I have described this aspect of relativism in my *Farewell to Reason* (London and New York: Verso, 1987), chap. 10.

*5. Humans as
Sculptors of Reality*

ACCORDING TO THE FIRST ASSUMPTION, our ways of thinking and speaking are products of idiosyncratic historical developments. Common sense and science both conceal this situation. For example, they say (second assumption) that atoms existed long before they were found. This explains why the projection received a response, but overlooks that vastly different projections did not remain unanswered.

A better way of telling the story is the following. Scientists, being equipped with a complex organism and embedded in constantly changing physical and social surroundings, used ideas and actions (and, much later, equipment up to and including industrial complexes such as CERN) to *manufacture,* first, metaphysical atoms, then, crude physical atoms, and, finally, complex systems of elementary particles out of a material that did not contain these elements but could be shaped into them. Scientists, according to this account, are sculptors of reality—but sculptors in a special sense. They not merely *act causally* upon the world (though they do that, too, and they have to if they want to "discover" new entities); they also *create semantic conditions* engendering strong inferences from known effects to novel projections and, conversely, from the projections to testable effects. We have here the same dichotomy of descriptions which Bohr introduced in his analysis of the case of Einstein, Podolsky, and Rosen.[28] Every individual, group, and culture tries to arrive at an equilibrium between the entities it posits and leading beliefs, needs, expectations, and ways of arguing. The separability assumption arises in special cases (traditions, cultures); it is not a condition (to be) satisfied by all, and it certainly is not a sound basis for epistemology. Altogether, the dichotomy subjective/objective and the corresponding dichotomy between descriptions and constructions are much too naïve to guide our ideas about the nature and the implications of knowledge claims.

28. Cf. the reprint in J. A. Wheeler and W. H. Zurek, eds., *Quantum Theory and Measurement* (Princeton: Princeton University Press, 1983), 42. The present essay is firmly based on Bohr's ideas. Reading the epilogue of Paul Hoyningen's *Die Wissenschaftsphilosophie Thomas Kuhns* (Braunschweig: Vieweg, 1989), I also realize that its ideas are very similar to, and almost identical with, Kuhn's as yet unpublished, later philosophy. I asked Hoyningen how he would explain this preestablished harmony (when writing my paper I was not familiar with Kuhn's later philosophy). His answer—"Reasonable people think along the same lines"—seems entirely acceptable.

I do not assert that any combined causal-semantic action will lead to a well-articulated and livable world. The material humans (and, for that matter, also dogs and monkeys) face must be approached in the right way. It *offers resistance;* some constructions (some incipient cultures—cargo cults, for example) find no point of attack in it and simply collapse. On the other hand, *this material is more pliable than is commonly assumed.* Molding it in one way (history of technology leading up to a technologically streamlined environment and large research cities such as CERN), we get elementary particles; proceeding in another, we get a nature that is alive and full of Gods. Even the "discovery" of America, which I used to support the separability assumption, allowed some leeway, as is shown by Edmondo O'Gorman's fascinating study, *The Invention of America.*[29] Science certainly is not the only source of reliable ontological information.

It is important to read these statements in the right way. They are not the sketch of a new theory of knowledge which explains the relation between humans and the world and provides a philosophical grounding for whatever discoveries are being made. Taking the historical character of knowledge seriously means rejecting any such attempt. We can describe the results we have obtained (though the description will always be fatally incomplete), we can comment on some similarities and differences that have come to our attention, and we can even try to explain what we found in the course of a particular approach "from the inside," i.e., using the practical and conceptual means provided by the approach (the theory of evolution, evolutionary epistemology, and modern cosmology belong in this category). We can tell many interesting *stories.* We cannot explain, however, how the chosen approach is related to the world and why it is successful, in terms of the world. This would mean knowing the results of all possible approaches or, what amounts to the same, we would know the history of the world before the world has come to an end.

And yet we cannot do without scientific know-how. Our world has been transformed by the material, spiritual, and intellectual impact of science and science-based technologies. Its reaction to the transformation (and a strange reaction it is!) is that we are stuck in a scientific environment. We need scientists, engineers, scientifically inclined philoso-

29. O'Gorman, *The Invention of America* (Bloomington: Indiana University Press, 1961).

phers, sociologists, etc., to deal with the consequences. My point is that these consequences are not grounded in an "objective" nature, but come from a complicated interplay between an unknown and relatively pliable material and researchers who affect and are affected and changed by the material which, after all, is the material from which they have been shaped. It is not therefore easier to remove the results. The "subjective" side of knowledge, being inextricably intertwined with its material manifestations, cannot be just blown away. Far from merely stating what is already there, it created conditions of existence, a world corresponding to these conditions and a life that is adapted to this world; all three now support or "establish" the conjectures that led to them. Still, a look at history shows that this world is not a static world populated by thinking (and publishing) ants who, crawling all over its crevices, gradually discover its features without affecting them in any way. It is a dynamical and multifaceted Being which influences and reflects the activity of its explorers. It was once full of Gods; it then became a drab material world; and it can be changed again, if its inhabitants have the determination, the intelligence, and the heart to take the necessary steps.

# {2} HAS THE SCIENTIFIC VIEW OF THE WORLD A SPECIAL STATUS COMPARED WITH OTHER VIEWS?

D ENNIS DIEKS[1] SKETCHES a framework which, he says, has guided the work of many physicists. He implies that the remaining conflicts are a purely philosophical phenomenon. Being fond of quarrels philosophers have split into schools. There are now empiricists, positivists, rationalists, anarchists, realists, apriorists, pragmatists, and they all have different views about the nature of science. Scientists, on the other hand, collaborate. Collaboration creates uniformity and, with it, a single way of looking at things: it *does* make sense to ask about the status of *the* scientific worldview.

In contrast I want to argue that scientists are as contentious as philosophers. But while philosophers merely talk, scientists act on their convictions; scientists from different areas use different procedures and construct their theories in different ways. Moreover, they often succeed: the worldviews we find in the sciences have empirical substance. *This is a fact, not a philosophical position.* I shall explain it by considering the following four questions:

1. What is *the* scientific view of the world and is there a single such view?

Published in Jan Hilgevoord, ed., *Physics and Our View of the World* (Cambridge: Cambridge University Press, 1994), 135–48, © Cambridge University Press 1994. Reprinted with kind permission of Cambridge University Press. The second page of this essay overlaps with part II, essay 8, section 4. One page toward the end of this essay overlaps with part I, chapter 3, section 1; with part II, essay 4, section 1; and with part II, essay 5.

1. Dennis Dieks, "The Scientific View of the World: Introduction," in *Physics and Our View of the World*, ed. Jan Hilgevoord (Cambridge: Cambridge University Press, 1994), 61–78.

2. Assuming there is a single scientific worldview—*for whom* is it supposed to be special?

3. What kind of *status* are we talking about? Popularity? Practical advantages? Truth?

4. What "other views" are being considered?

My answer to the *first question* is that the wide divergence of individuals, schools, historical periods, and entire sciences makes it difficult to identify comprehensive principles either of method, or of fact.

In the domain of *method* we have scientists like Luria who want to tie research to events permitting "strong inferences," "predictions that will be strongly supported and sharply rejected by a clear-cut experimental step."[2]

According to Luria and Delbrück the experiments,[3] which showed that the resistance of bacteria to phage invasion is a result of environment-independent mutations and not of an adaptation to the environment, had precisely this character. There was a simple prediction; the prediction could be tested in a straightforward way yielding a decisive and unambiguous result. (The result refuted Lamarckism, which was popular among bacteriologists but practically extinct elsewhere—a first indication of the complexity of science.)

Scientists of Luria's inclination show a considerable "lack of enthusiasm in the 'big problems' of the Universe or of the early Earth or in the concentration of carbon dioxide in the upper atmosphere,"[4] all subjects that are "loaded with weak inferences."[5] In a way they are continuing the Aristotelian approach, which demands close contact with experience and objects to following a plausible idea to the bitter end.[6]

However this was precisely the procedure adopted by Einstein, by researchers in celestial mechanics between Newton and Poincaré (stability of the planetary system), by the proponents of atomism and, later, of the kinetic theory, by Heisenberg during the initial stages of matrix mechanics (when it seemed to clash with the tracks produced in the Wilson

2. S. E. Luria, *A Slot Machine, a Broken Test Tube* (New York: Harper and Row, 1984), 115.

3. S. E. Luria and M. Delbrück, "Mutations of Bacteria from Virus Sensitivity to Virus Resistance," *Genetics* 28 (1943): 491–511.

4. Luria, *A Slot Machine*, 119.

5. Ibid.

6. Aristotle, *De caelo*, 293a25 ff.

chamber), and by almost all cosmologists. Einstein's first cosmological paper is a purely theoretical exercise containing not a single astronomical constant. The subject of cosmology itself for a long time found few supporters among physicists. Hubble, the observer, was respected, the rest had a hard time:

> Journals accepted papers from observers, giving them only the most cursory refereeing whereas our own papers always had a stiff passage, to a point where one became quite worn out with explaining points of mathematics, physics, fact and logic to the obtuse minds who constitute the mysterious anonymous class of referees, doing their work, like owls, in the darkness of the night.[7]

"Is it not really strange," asks Einstein,[8] "that human beings are normally deaf to the strongest argument while they are always inclined to overestimate measuring accuracies?" But just such an "overestimating of measuring accuracies" is the rule in epidemiology, demography, genetics, spectroscopy, and other subjects. The variety increases when we move into sociology or cultural anthropology, where a compromise has to be found between the effects of personal contact, the pressing needs of a region, and the idea of objectivity. Robert Chambers, the inventor of the method of rapid rural appraisal, writes as follows:

> To hear a seminar at the university about modes of production in the morning and then attend a meeting in a government office about agricultural extension in the afternoon leaves a schizoid feeling. One might not know that both referred to the same small farmers and might doubt whether either discussion has anything to contribute to the other.[9]

Methods are not restricted to particular areas. Luria's requirements seem to be tied to laboratory science and, more especially, to bacteriology; however they also turn up in astrophysics and cosmology. For example, they were used by Heber Curtis in 1921, in his "grand debate" with Harlow Shapley. (Curtis doubted that stellar features could be as readily generalized as was assumed by Shapley—and he was right, especially in the decisive case of the Cepheides.) They guided the great Armenian astrophysicist Viktor Ambarzumjan and they are now being applied by

7. F. Hoyle, "Steady State Cosmology," in *Cosmology and Astrophysics,* ed. Y. Terzian and S. M. Bilson (Ithaca: Cornell University Press, 1982), 21.

8. *The Born-Einstein Letters* (New York: Walter, 1971), 192.

9. Chambers, *Rural Development* (London: Longman, 1986), 29.

Halton Arp, Margaret Geller, and their collaborators. In Prandtl's lectures we read:

> The great growth in technical achievement which began in the nineteenth century left scientific knowledge far behind. The multitudinous problems of practice could not be answered by the hydrodynamics of Euler; they could not even be discussed. This was chiefly because, starting from Euler's equations of motion, the science had become more and more a purely academic analysis of the hypothetical frictionless "ideal fluid." This theoretical development is associated with the names of Helmholtz, Kelvin, Lamb and Rayleigh.
>
> The analytical results obtained by means of this so called "classical hydrodynamics" virtually do not agree at all with the practical phenomena . . . Therefore the engineers . . . put their trust in a mass of empirical data collectively known as the "science of hydraulics," a branch of knowledge which grew more and more unlike hydrodynamics.[10]

According to Prandtl we have a disorderly collection of facts on the one side, sets of theories starting from simple but counterfactual assumptions on the other and no connection between the two. More recently the axiomatic approach in quantum mechanics and especially in quantum field theory was compared by clinical observers to the Shakers,

> a religious sect of New England who built solid barns and led celibate lives, a non-scientific equivalent of proving rigorous theorems and calculating no cross sections.[11]

Yet in quantum mechanics this apparently useless activity has led to a more coherent and far more satisfactory codification of the facts than had been achieved before, while in hydrodynamics "physical common sense" occasionally turned out to be less accurate than the results of rigorous proofs based on wildly unrealistic assumptions. An early example is Maxwell's calculation of the viscosity of gases. For Maxwell this was an exercise in theoretical mechanics, an extension of his work on the rings of Saturn. Neither he nor his contemporaries believed the result—that viscosity remains constant over a wide range of density—and there

---

10. L. Prandtl and O. G. Tietjens, *Fundamentals of Hydro- and Aerodynamics* (New York: Dover, 1957), 3.

11. R. F. Streater and A. S. Wightman, *PCT, Spin, Statistics and all that* (New York: W. A. Benjamin, 1964), 1.

was contrary evidence. Yet more precise measurements turned the apparent failure into a striking success.[12]

Meanwhile the situation has changed in favor of theory. In the 1960s and 1970s, when science was still admired, theory got the upper hand, at universities where it increasingly replaced professional skills, and in special subjects such as biology or chemistry where earlier morphological and substance-related research was replaced by a study of molecules. In cosmology a firm belief in the Big Bang now tended to devalue observations that clashed with it. C. Burbidge writes:

> Such observations are delayed at the refereeing stage as long as possible with the hope that the author will give up. If this does not occur and they are published the second line of defence is to ignore them. If they give rise to some comment, the best approach is to argue simply that they are hopelessly wrong and then, if all else fails, an observer may be threatened with loss of telescope time until he changes his program.[13]

These and similar examples show that science contains different trends with different research philosophies. One trend requires that scientists stick closely to the facts, design experiments that clearly establish the one or the other of two conflicting alternatives, and avoid far-reaching speculations. One might call it the Aristotelian trend. Another trend encourages speculation and is ready to accept theories that are related to the facts in an indirect and highly complex way. Let us call this the Platonic trend. The existence of different trends within a comprehensive venture is not surprising. On the contrary, it would be strange if large groups of passionate and imaginative people, who despise authority and make criticism a guide to research, subscribed to a single point of

12. For quantum mechanics see Hans Primas, *Chemistry, Quantum Mechanics, and Reductionism* (Berlin: Springer, 1981), secs. 4.1 and 4.2. Maxwell's calculations are reproduced in *The Scientific Papers of James Clerk Maxwell*, ed. W. D. Niven (1890; reprint, New York: Dover, 1965), 377 ff. The conclusion is stated on p. 391: "A remarkable result here presented to us . . . is that if this explanation of gaseous friction be true, the coefficient of friction is independent of the density. Such a consequence of mathematical theory is very startling, and the only experiment I have met with on the subject does not seem to confirm it." For examples from hydrodynamics see G. Birkhoff, *Hydrodynamics* (New York: Dover, 1955), secs. 20 and 21.

13. C. Burbidge, "Problems of Cosmogony and Cosmology," in F. Bertola, J. W. Sulentic, D. F. Madore, eds., *New Ideas in Astronomy* (Cambridge: Cambridge University Press, 1988), 229.

view. What is surprising is that almost all the trends that developed within the sciences, Aristotelianism and an extreme Platonism included, produced results, not only in special domains, but everywhere; there exist highly theoretical branches of biology and highly empirical parts of astrophysics. The world is a complex and many-sided thing.

So far I have been talking about procedures, or methods. Now methods that are used as a matter of habit, without any thought about the reasons behind them, are often tied to metaphysical beliefs. Aristotelians assume that humans are in harmony with the Universe; observation and truth are closely related. For Platonists humans are deceived in many ways. It needs abstract thought to get in touch with reality. Adding empirical success to these and other trends we arrive at the result that *science contains many different and yet empirically acceptable worldviews, each one containing its own metaphysical background.* Johann Theodore Merz[14] describes the situation in the nineteenth century. He discusses the following views. *The astronomical view* rested on mathematical refinements of action-at-a-distance laws and was extended (by Coulomb, Neumann, Ampère, and others) to electricity and magnetism. Laplace's theory of capillarity was an outstanding achievement of this approach. (In the eighteenth century Benjamin Rush had applied the view to medicine thus creating a unified but deadly medical system.) *The atomic view* played an important role in chemical research (e.g., stereochemistry) but was also opposed by chemists, for empirical as well as for methodological reasons. *The kinetic and mechanical view* employed atoms in the area of heat and electric phenomena. For some scientists (but by no means for all) atomism was the foundation of everything. *The physical view* tried to achieve universality in a different way, on the basis of general notions such as the notion of energy. It could be connected with the kinetic view, but often was not. Physicians, physiologists, and chemists like Mayer, Helmholtz, du Bois-Reymond, and, in the practical area, Liebig, were outstanding representatives in the later nineteenth century while Ostwald, Mach, and Duhem extended it into the twentieth. *The phenomenological view* (not mentioned by Merz) is related to the physical view, but less general. It was adopted by scientists like Lamé who found that it provided a more direct way to theory (elasticity, in the case of Lamé)

14. Johann Theodore Merz, *A History of European Thought in the Nineteenth Century* (1903; reprint, New York: Dover, 1965).

than the use of atomic models. Starting his description of *the morphological view* Merz writes:

> The different aspects of nature which I have reviewed in the foregoing chapters and the various sciences which have been elaborated by their aid, comprise what may appropriately be termed the abstract study of natural objects and phenomena. Though all the methods of reasoning with which we have so far become acquainted originated primarily through observation and the reflection over things natural, they have this in common that they— for the purpose of examination—remove their objects out of the position and surroundings which nature has assigned to them: that they *abstract* them. This process of abstraction is either literally a process of removal from one place to another, from the great work- and storehouse of nature herself to the small workroom, the laboratory of the experimenter; or—where such removal is not possible—the process is carried out merely in the realm of contemplation; one or two special properties are noted and described, whilst the number of collateral data are for the moment disregarded. [A third method, not developed at the time, is the creation of "unnatural" conditions and, thereby, the production of "unnatural" phenomena.]
>
> There is, moreover, in addition to the aspect of convenience, one very powerful inducement for scientific workers to persevere in their process of abstraction . . . This is the practical usefulness of such researches in the arts and industries . . . The wants and creations of artificial life have thus proved the greatest incentives to the abstract and artificial treatment of natural objects and processes for which the chemical and electrical laboratories with the calculating room of the mathematician on the one side and the workshop and factory on the other, have in the course of the century become so renowned . . .
>
> There is, however, in the human mind an opposite interest which fortunately counteracts to a considerable extent the one-sided working of the spirit of abstraction in science . . . This is the genuine love of nature, the consciousness that we lose all power if, to any great extent, we sever or weaken that connection which ties us to the world as it is—to things real and natural: it finds its expression in the ancient legend of the mighty giant who derived all his strength from his mother earth and collapsed if severed from her . . . In the study of natural objects we meet (therefore) with a class of students who are attracted by things as they are . . . [Their] sciences are the truly descriptive sciences, in opposition to the abstract ones.[15]

15. Ibid., 2:200 f.

I have quoted this description at length for it shows very clearly how different interests leading to different procedures collect evidence which can then congeal into an empirically founded worldview. Finally, Merz mentions the genetic view, the psychophysical view, the vitalistic view, the statistical view together with their procedures and their findings.

What can a single comprehensive "worldview of science" offer under such circumstances?

It can offer a survey, a *list* similar to the list given by Merz, enumerating the achievements and drawbacks of the various approaches as well as the clashes between them, and it can identify science with this complex and somewhat scattered war on many fronts. Alternatively it can put one view on top and *subordinate* the others to it, either by pseudoderivations, or by declaring them to be meaningless. Reductionists love to proceed in this fashion. Or it can disregard the differences and present a "paste job" where each particular view and the results it has achieved is smoothly connected with the rest, thus producing an impressive and coherent edifice—*the* scientific worldview.

Expressing it differently we may say that the assumption of a single coherent worldview that underlies all of science is either a metaphysical hypothesis trying to anticipate a future unity, or a pedagogical fake; or it is an attempt to show, by a judicious up- and downgrading of disciplines, that a synthesis has already been achieved. This is how fans of uniformity proceeded in the past (cf. Plato's list of subjects in chapter seven of his *Republic*), these are the ways that are still being used today. A more realistic account, however, would point out that

> [t]here is no simple "scientific" map of reality—or if there were, it would be much too complicated and unwieldy to be grasped or used by anyone. But there are many different maps of reality, from a variety of scientific viewpoints.[16]

You may object that we live in the twentieth century, not in the nineteenth, and that unifications which seemed impossible then have been achieved by now. Examples are statistical thermodynamics, molecular biology, quantum chemistry, and superstrings. These are indeed flourishing subjects, but they have not produced the unity the phrase "*the* scientific view of the world" insinuates. In fact, the situation is not very dif-

---

16. John Ziman, *Teaching and Learning about Science and Society* (Cambridge: Cambridge University Press, 1980), 19.

ferent from what Merz had noticed in the nineteenth century. Truesdell and others continue the physical approach. Prandtl maligned Euler; Truesdell praises him for his rigorous procedure. Morphology, though given a low status by some and declared to be dead by others, has been revived by ecologists and by Lorenz's study of animal behavior (which added forms of motion to the older static forms); it always played an important role in galactic research (Hubble's classification) and astrophysics (Hertzsprung-Russell diagram). Having been in the doghouse cosmology is now being courted by high-energy physicists but clashes with the philosophy of complementarity accepted by the same group. Commenting on the problem M. Kafatos and R. Nadeu write:

> The essential requirement of the Copenhagen interpretation that the experimental setup must be taken into account when making observations is seldom met in observations with cosmological import [though such observations rely on light, the paradigm case of complementarity].[17]

Moreover, the observations of H. Arp, M. Geller, and others have thrown doubt on the homogeneity assumption which plays a central role in cosmology. We have a rabid materialism in some parts (molecular biology, for example), a modest to radical subjectivism in others (some versions of quantum measurement, anthropic principle). There are many fascinating results, speculations, attempts at interpretation, and it is certainly worth knowing them. But pasting them together into a single coherent "scientific" worldview, a procedure which has the blessings even of the pope[18]—this is going too far. After all—who can say that the world which so strenuously resists unification really is as educators and metaphysicians want it to be—tidy, uniform, the same everywhere? Besides, a "paste job" eliminates precisely those conflicts that kept science going in the past and will continue inspiring its practitioners if preserved.

---

17. M. Kafatos and R. Nadeu, "Complementarity and Cosmology," in M. Kafatos, ed., *Bell's Theorem, Quantum Theory, and Conceptions of the Universe* (Dordrecht: Kluwer Academic Publishers, 1989), 263; see also Wheeler's thought experiment involving a quasar situated behind a galaxy.

18. See Pope John Paul II's message on the occasion of the three hundredth anniversary of Newton's *Principia*, published in *John Paul II on Science and Religion*, ed. R. J. Russell, W. R. Stoeger S.J., and G. V. Coyne (Vatican City: Vatican Observatory Publications, 1990), esp. M6 ff.

I NOW COME TO THE *SECOND QUESTION:* special status—for whom? I do not want to argue that considerations of status (truth, reality) are necessarily relativistic. But social considerations do play a role and they have occasionally advanced the cause of science. Consider the following example (which involves far-reaching interpretations from slender data).

Like modern scientists the Ionian natural philosophers (Thales, Anaximander, Anaximenes) looked for simple principles behind the variety of phenomena. Today the principles sought are theories or laws. In ancient Greece they were substances. According to Thales water was the basis of everything. This is not as implausible as it sounds. The Greeks, who "lived around the Mediterranean like frogs around a pond," could *see* how water transformed, first, into mist, then into air and perhaps even into fire (lightning). Frozen water was solid (earth) and, besides, water was needed everywhere, to sustain life. Using a symmetry principle Anaximander objected that fire, earth, and air seemed to be as important as water, which means that the basic substance had to be different from all elements, though capable of turning into them under special circumstances. Anaximander called it *apeiron*—the unlimited. Parmenides then pointed out that Being was still more fundamental (water *is*, fire *is*, *apeiron is*—they are all forms of Being). What can be said about Being? That it is and that not-Being is not. Note that the statement BEING IS (*estin* in the Greek of Parmenides) was the first explicit conservation principle of the West: it asserted the conservation of Being. Accepting this argument we can infer that there is no change: the only possible change is into not-Being, not-Being does not exist, hence there is no change. What about difference? The only possible difference is between Being and not-Being, not-Being does not exist, hence Being is everywhere the same. But don't we perceive change and difference? Yes, we do, which shows that change and difference are appearances, chimeras. Reality does not change. This was the first and most radical (Western) theory of knowledge. It is not entirely ridiculous: nineteenth-century science up to and including Einstein also devalued change. Hermann Weyl writes:

> The relativistic world simply *is*, it does not *happen*. Only to the gaze of my consciousness, crawling upward along the lifeline of my body, does a section of this world come to life as a fleeting image in space which continuously changes in time.[19]

19. Hermann Weyl, *Philosophy of Mathematics and Natural Science* (Princeton: Princeton University Press, 1949), 116.

Ancient atomism can be seen as an attempt to shorten the distance between basic physics (BEING IS) and common sense. Leukippos and Democritos retained one part of Parmenides' theory (atoms are tiny fragments of Parmenidean Being) and rejected another (not-Being exists and it is identical with space). There was no refutation—after all, Parmenides had proved that change was apparent and you cannot refute a theory about real things by confronting it with appearances. The aim was, rather, to adapt physics to certain social tendencies (such as the tendency to regard change as something very important). In Aristotle this aim is stated explicitly: real is what plays a basic role in the life we want to lead. The second question now amounts to this: are we prepared to view ourselves in the manner suggested by scientists or do we prefer to make personal contact, friendship, etc. the measure of our nature? Note that what is needed here is a personal (social) decision, not a scientific argument. Instrumentalists explain how we can have our cake, i.e., how we can continue holding familiar views (religious views, for example), and eat it too (i.e., profit from the practical results of the sciences).

*THIRD QUESTION:* what do we mean by status? *Popularity,* i.e., familiarity with major results and the admission that they are important, would be one measure. Now it is true that despite periodic swings toward the sciences and away from them the sciences are still in high repute with the general public. Or, to be more precise, not the sciences are, but a mythical monster "science" (in the singular—in German it sounds even more impressive DIE WISSENSCHAFT). For what the so-called "educated public" seems to assume is that the achievements they read about in the educational pages of their newspapers and the threats they seem to perceive come from a single source and are produced by a uniform procedure. I have tried to argue that scientific practice is much more diverse. Adding that 10 percent of all average Germans still believe in a stable Earth, that a third of all adults in the USA believe everything in the Bible to be literally true,[20] that a promotional page for a new journal, *Public Understanding of Science,* contains this passage—

> Annual polls show again and again a yawning gap between public attitudes and scientific advances. In a Gallup poll commissioned this summer [1991] a pic-

---

20. *International Herald Tribune,* December 21–25, 1991, p. 4.

ture emerges of a population which not only confesses ignorance but also a substantial lack of concern about the giant discoveries transforming everyday life.

—and that the "population" referred to is the Western middle class, not Bolivian peasants (for example) I conclude that popularity can hardly be regarded as a measure of excellence.

What about *practical advantages?* The answer is that "science" sometimes works and sometimes doesn't but that it is still sufficiently mobile to turn disaster into triumph. *It can do that because it is not tied to particular methods or worldviews.* The fact that an approach or a subject (economics, for example) is "scientific" according to some abstract criterion is therefore no guarantee that it will succeed. *Each case must be judged separately,* especially today, when the inflation of the sciences has added some rather doubtful activities to what used to be a sober enterprise.[21] The question of *truth,* finally, remains unresolved. Love of Truth is one of the strongest motives for replacing what really happens by a streamlined account or, to express it in a less polite manner—love of truth is one of the strongest motives for deceiving oneself and others. Besides, the quantum theory seems to show, in the precise manner beloved by the admirers of science, that reality is either one, which means there are no observers and no things observed, or it is many, including theoreticians, experimenters, and the things they find, in which case what is found does not exist in itself but depends on the procedure chosen.

I NOW COME TO THE *LAST QUESTION:* what are the other views that are being considered? In my public talk I quoted a passage from E. O. Wilson.[22] It runs as follows:

> . . . religion . . . will endure for a long time as a vital force in society. Like the mythical giant Antaeus who drew energy from his mother, the earth, religion cannot be defeated by those who may cast it down. The spiritual weakness of scientific naturalism is due to the fact that it has no such pri-

21. This was realized by government advisers after the postwar euphoria had worn off. "The idea of a comprehensive science policy was gradually abandoned. It was realized that science was not one but many enterprises and that there could be no single policy for the support of all of them." (Joseph ben-David, *Scientific Growth* [Berkeley and Los Angeles: University of California Press, 1991], 525.)

22. E. O. Wilson, *On Human Nature* (Cambridge: Harvard University Press, 1978), 192 f.

mal source of power . . . So the time has come to ask: does a way exist to di-vert the power of religion into the services of the great new enterprise?

For Wilson the main feature of the alternatives is that they have *power.* I regard this as a somewhat narrow characterization. Worldviews also an-swer questions about origins and purposes which sooner or later arise in almost every human being. Answers to these questions were available to Kepler and Newton and were used by them in their research; they are no longer available today, at least not within the sciences. They are part of nonscientific worldviews which therefore have much to offer, also to sci-entists. When Western civilization invaded the Near and Far East and what is now called the Third World it imposed its own ideas of a proper environment and a rewarding life. Doing this, it disrupted delicate pat-terns of adaptation and created problems that had not existed before. Hu-man decency and an appreciation of the many ways in which humans can live with nature have prompted agents of development and public health to think in more complex or, as some would say, more "relativistic" ways—which is in complete agreement with the pluralism of science itself.

To SUM UP: there is no "scientific worldview" just as there is no uniform enterprise "science"—except in the minds of metaphysicians, school-masters, and scientists blinded by the achievements of their own partic-ular niche. Still, there are many things we can learn from the sciences. But we can also learn from the humanities, from religion, and from the remnants of ancient traditions that survived the onslaught of Western civilization. No area is unified and perfect, few areas are repulsive and completely without merit. There is no objective principle that could di-rect us away from the supermarket "religion" or the supermarket "art" to-ward the more modern, and much more expensive, supermarket "sci-ence." Besides, the search for such guidance would be in conflict with the idea of individual responsibility which allegedly is an important ingredient of a "rational" or scientific age. It shows fear, indecision, a yearning for authority, and a disregard for the new opportunities that now exist: we can build worldviews on the basis of a personal choice and thus unite, for ourselves and for our friends, what was separated by the chauvinism of special groups.

On the other hand, we can agree that in a world full of scientific products scientists may be given a special status just as henchmen and generals had a special status at times of social disorder or priests when being a citizen coincided with being the member of a church. We can also agree that appealing to a chimera can have important political consequences. In 1854 Commander Perry, using force, opened the ports of Hakodate and Shimoda to American ships for supply and trade. This event demonstrated the military inferiority of Japan. The members of the Japanese enlightenment of the early 1870s, Fukuzawa among them, now reasoned as follows: Japan can keep its independence only if it becomes stronger. It can become stronger only with the help of science. It will use science effectively only if it does not just practice science but also believes in the underlying ideology. To many traditional Japanese this ideology— "the" scientific worldview—was barbaric. But, so the followers of Fukuzawa argued, it was necessary to adopt barbaric ways, to regard them as advanced, to introduce the whole of Western civilization in order to survive.[23] Having been thus prepared, Japanese scientists soon branched out as their Western colleagues had done before and falsified the uniform ideology that had started the development. The lesson I draw from this sequence of events is that a uniform "scientific view of the world" may be useful *for people doing science*—it gives them motivation without tying them down. It is like a flag. Though presenting a single pattern it makes people do many different things. However, *it is a disaster for outsiders* (philosophers, fly-by-night mystics, prophets of a New Age, the "educated public"), who, being undisturbed by the complexities of research, are liable to fall for the most simpleminded and most vapid tale.

23. Details in Carmen Blacker, *The Japanese Enlightenment* (Cambridge: Cambridge University Press, 1969). For the political background see Richard Storry, *A History of Modern Japan* (Harmondsworth: Penguin, 1982), chaps. 3 and 4.

⟨3⟩ QUANTUM THEORY AND

OUR VIEW OF THE WORLD

W
HEN I WAS A STUDENT in Vienna, in the late 1940s, we had
three physicists who were known to a wider public: Karl
Przibram, Felix Ehrenhaft, and Hans Thirring. Przibram
was an experimentalist, a pupil of J. J. Thomson, whom he often mentioned with reverence. Philosophers of science know him as the editor of
a correspondence on wave mechanics between Schrödinger, Lorentz,
Planck, and Einstein. He was the brother of Hans Przibram, the biologist, and, I believe, the uncle of the neurophysiologist Karl Przibram. He
talked with a subdued voice and wrote tiny equations on the blackboard.
Occasionally his lectures were interrupted by shouting, laughing, and
trampling from below; that was Ehrenhaft's audience.

Ehrenhaft had been professor of theoretical and experimental physics
in Vienna. He left when the Nazis came; he returned in 1947. By that
time many physicists regarded him as a charlatan. He had produced and
kept producing evidence for subelectrons, magnetic monopoles of meso-
scopic size, and magnetolysis, and he held that the inertial path was a
spiral, not a geodesic. His attitude toward theory was identical with that
of Lenard and Stark, whom he often mentioned with approval. He challenged us to criticize him and laughed when he realized how strongly we
believed in the excellence of, say, Maxwell's equations without having
calculated and tested specific effects. During a summer school in Alp-
bach he set up his experiments in a little farmhouse and invited everyone

Published in Jan Hilgevoord, ed., *Physics and Our View of the World* (Cambridge:
Cambridge University Press, 1994), 149–67, © Cambridge University Press 1994. Reprinted with kind permission of Cambridge University Press. This essay is based on a
lecture delivered at the Erasmus Ascension Symposium in Leiden, April 1992, first published in *Stroom* 28 (December 1992): 19–24. Section 3 of this essay has some overlap with
part I, chapter 3, section 1; section 4 has some overlap with part I, chapter 3, section 2;
and with part II, essay 5.

to have a look. Leon Rosenfeld was there; so was Maurice Pryce, one of the most abrasive physicists of his generation. They went in; when they reappeared they looked as if they had seen something obscene. However, all they could say was "obviously a *Dreckeffekt.*" Afterward, in Ehrenhaft's lecture, Rosenfeld and Pryce sat in the front row. Having described his experiments Ehrenhaft went up to them and exclaimed: "Was können Sie sagen mit allen Ihren schönen Theorien? Nichts können Sie sagen. Still müssen Sie sein. Sitzen müssen Sie bleiben." (What can you say with all your fine theories? You can say nothing. You must be silent. You have to remain seated.) (The German *sitzenbleiben* means [1] to remain seated and [2] to repeat a school class.) And, indeed, Rosenfeld and Pryce, so eloquent on other occasions, did not say a single word. Ehrenhaft may not have been mainstream. But he made us think—more than many mainstream scientists before and after him.

The third physicist was the theoretician Hans Thirring, the father of Walter Thirring and discoverer of the (general-relativistic) effect of a rotating shell on objects inside. Thirring was a pacifist and a friend of Einstein, Freud, and other disreputable people. He was sacked when Austria became part of the Reich. When he returned he had completed a huge manuscript on the psychological foundations of world peace. "This is important," he said; "physics is not." Later, as a member of parliament he suggested that Austria abandon its army and have its borders guaranteed by the neighboring powers; needless to say this suggestion did not get anywhere. Thirring had strong convictions; yet he never lost his sense of humor. For him the events in Germany and the war afterward were signs of human folly, not of evil incarnate. He was a rare combination, a committed humanitarian *and* a skeptic.

Przibram, Ehrenhaft, and Thirring differed in many ways. They were united in their belief that worldviews are dangerous things and that physics should do without them. The belief had been held before—by Mach, Boltzmann, by Franz Exner and his group, and by the members of the Vienna Circle (a group of logical positivists centered on Vienna University in the 1920s and 1930s). Only a few of these thinkers realized that they were guided by a rival worldview which, though humanitarian in intention, disregarded important features of human existence.

Today the situation is much less uniform. Many scientists now raise the question of religion; others are looking for ways of making science as powerful as religion used to be. I shall try to clarify this development by considering to what extent worldviews affected and perhaps even created scientific knowledge.

*1. The Nature*     IN HIS ESSAY *On Human Nature*, E. O. Wilson,
*of Worldviews*     whom some regard as the father of sociobiology,
             writes about religion as follows.

> [R]eligion . . . will endure for a long time as a vital force in society. Like the mythical giant Antaeus who drew energy from his mother, the earth, religion cannot be defeated by those who may cast it down. The spiritual weakness of scientific naturalism is due to the fact that it has no such primal source of power . . . So the time has come to ask: does a way exist to divert the power of religion into the services of the great new enterprise?[1]

According to Wilson there is religion, and there is this "great new enterprise," scientific naturalism. Both look at the world in a rather general way. But while scientific naturalism provides information and makes practical suggestions, religion in addition is a "primal [i.e., a nonpolitical] source of power."

Jacques Monod says the same, though with greater understanding of the problems involved.

> Cold and austere, proposing no explanation but imposing an ascetic renunciation of all other spiritual fare [the idea that objective knowledge is the only authentic source of truth] was not of a kind to allay anxiety but aggravated it instead. By a single stroke it claimed to sweep away the tradition of a hundred thousand years, which had become one with human nature itself. It wrote an end to the ancient animist covenant between man and nature, leaving nothing in place of that precious bond but an anxious quest in a frozen universe of solitude. With nothing to recommend it but a certain puritan arrogance, how could such an idea win acceptance? It did not; it still has not. It has however commanded recognition; but that it did solely because of its prodigious power of performance.[2]

According to Monod objectivism—and by this Monod means knowledge not involving aims and purposes—has great achievements to its credit. However, it provides only part of what is needed for a full life. As-

1. E. O. Wilson, *On Human Nature* (Cambridge: Harvard University Press, 1978), 192 f.
2. Jacques Monod, *Chance and Necessity* (New York: Vintage Books, 1972), 170 (passage in brackets from 169).

serting that there is nothing more to be had it produced a meaningless world. Or to repeat a much commented-upon statement by Steven Weinberg—the more the universe seems comprehensible, the more it seems pointless.

My third quotation is from a letter Wolfgang Pauli wrote to his colleague Markus Fierz. I shall say more about Wolfgang Pauli below. Here I only present the quotation.

> Science wants to examine reality. This problem is closely connected with the other problem, namely, *the idea of reality*. When speaking about reality the layman usually assumes that he is talking about something that is obvious and known. However to me it is an important and very difficult task of our times to work at building a new idea of reality. This is also what I mean when emphasizing . . . that science and religion must in some way be related to each other.[3]

Pauli is looking for a point of view that pays attention to science but also goes beyond it.

Taking these three quotations as my guide I shall define a *worldview* as a collection of beliefs, attitudes, and assumptions that involves the whole person, not only the intellect, has some kind of coherence and universality, and imposes itself with a power far greater than the power of facts and fact-related theories.

*2. The Power of Worldviews*

BEING CONSTITUTED IN THIS MANNER worldviews have tremendous strength. They prevail despite the most obvious contrary evidence and they increase in vigor when meeting obstacles.[4] Cruel wars, deadly epi-

---

3. Wolfgang Pauli, letter to Markus Fierz (August 18, 1948), in C. A. Maier ed., *Wolfgang Pauli und C. G. Jung: Ein Briefwechsel 1932–1958* (Berlin: Springer, 1992).

4. Leon Festinger, *When Prophecy Fails* (1956; reprint, Harper and Row, 1988) gives examples of the power of worldviews and tries to explain why obstacles increase it. This property is very useful. New ideas are crude, unfinished, unfamiliar, and ill adapted to their natural and social surroundings. It is easy for opponents to prove their (empirical, logical, social) imperfection. Ideas need time to develop their advantages and strength to survive initial attacks. They must therefore appear in the form of worldviews, not of theories, and their defenders must disregard prima facie conflicts with logic, evidence, and accepted principles. Many scientists proceeded in this way (examples in Paul Feyerabend, "Was heisst das—wissenschaftlich sein?" in *Grenzprobleme der Wissenschaften*, ed. P. Feyerabend and Chr. Thomas [Zurich: Verlag der Fachvereine, 1985], 385 ff).

But the property is also quite dangerous—political history, the history of medical and

demics that killed people indiscriminately, natural catastrophes, floods, earthquakes, widespread famines could not overcome the belief in an all-powerful, just, and even benign creator god. Altogether it seems that people who are guided by worldviews are incapable of learning from experience.

For enlightened people this apparent irrationality is one of the strongest arguments against all forms of religion. What they fail to realize is that *the rise of the sciences depended on a blindness, or obstinacy, of exactly the same kind.* Surrounded by comets, new stars, plagues, strange geological shapes, unknown illnesses, irrational wars, biological malformations, meteors, oddities of weather, the leaders of Western science asserted the universal, "inexorable and immutable" character of the basic laws of Nature.[5] Early Chinese thinkers had taken the empirical variety at face value. They had favored diversification and had collected anomalies instead of trying to explain them away.[6] Aristotelians had emphasized the local character of regularities and insisted on a classification by multiple substances and corresponding accidents. Natural is what happens always or almost always, said Aristotle.[7] Scientists of an equally empirical bent, Tycho Brahe among them, regarded some anomalous events of their time as miracles or, as we might say today, they took cosmic idiosyncrasies seriously; others, like Kepler, ascribed them to the subjective reactions of the telluric soul while the great Newton, for empirical as well as for theological reasons, saw in them the finger of God.[8]

---

biological fashions provide numerous examples. It needs tact, wisdom, judgment to stop in time and thus to prevent disasters. Science would fail without a (nonformalizable) sense of the right balance between boldness and caution.

5. Galileo Galilei, letter to Benedetto Castelli (December 21, 1613).

6. On early Chinese views of the Universe see Fang Lizhi, "Notes on the Interface between Science and Religion," in *John Paul II on Science and Religion,* ed. R. J. Russell, W. R. Stoeger S.J., and G. V. Coyne (Vatican City: Vatican Observatory Publications, 1990). For Aristotle see Kurt Lewin, "Der Übergang vom Aristotelischen zum Galileischen Denken in Biologie und Psychologie," *Erkenntnis* 2(1931). For Kepler see Norbert Herz, *Kepler's Astrologie* (Vienna: Carl Gerold's Sohn, 1885), esp. 24 f.

7. Aristotle, *De partibus animalium* 663b27 ff.

8. Isaac Newton, *Opticks* (New York: Dover, 1979), query 31. Historians and scientists for a considerable time tried to keep Newton's theological and alchemical writings out of sight. How this was done is described in R. Popkin, *The Third Force in 17th Century Thought* (Leiden: E. J. Brill, 1992), 189–94. Newton, these scholars thought, was a scientist, his theology was an aberration, almost an obscenity that had no place in a balanced account of his life. How could a scientist of Newton's stature waste his time on nonsense like that? Today the question is different. How did it happen, asks Popkin, that one of the foremost anti-Trinitarian theologians of his age could be sidetracked into astronom-

Even instrumentalism, the doctrine, that is, that scientific theories do not describe reality but are instruments for the prediction of what can be observed, had a metaphysical or, as one might say, a worldview backing. According to Saint Thomas universals existed side by side with God, only in a different manner,[9] while Duns Scotus and William of Ockham objected that all we have are the results of God's acts of will.[10] We do not understand why these acts occurred, we do not know what acts will occur in the future, we can only observe their results, connect them and hope for the best. It needed a tremendous act of faith or, to express it in the terminology I am using here, it needed guidance by a powerful worldview not only to *assume* regularities where no regularities had yet been found, but *to work toward them* for centuries, and in the face of numerous failures.

---

ical and physical research? R. S. Westfall, *Never at Rest* (Cambridge: Cambridge University Press, 1980), 875 ff. gives the location of relevant manuscripts and literature, Frank E. Manuel, *The Religion of Isaac Newton* (Oxford: Clarendon Press, 1974), background and quotations. B. J. T. Dobbs, *The Janus Face of Genius* (Cambridge: Cambridge University Press, 1991), contains a fascinating and amply documented synthesis of Newton's religion, science, historical research, and alchemy and shows the impact his scientific discoveries had on his overall worldview.

Bacon had advised researchers to keep philosophy and revelation separated (*The Advancement of Learning*, ed. S. E. Creighton [New York: Willey, 1944], 5). This assumed two sources of knowledge, or two books, God's Words (the Bible) and God's Works (the Universe). The idea that Nature should be approached without prejudice was therefore in part religiously founded: you accept what God tells you, you do not impose your "interpretations" on it. Newton upheld the separation, most likely for political reasons. When he was president of the Royal Society "he banned anything remotely touching religion, even apologetics" (Manuel, *The Religion*, 30). He did make theological remarks, but he put them into scholia, queries, letters—not into his main argument. Altogether he thought and argued in each area (theology, science, alchemy) according to the standards of this area.

Leibniz's assumption (first paper of the *Leibniz-Clarke Correspondence*, ed. H. G. Alexander [Manchester: Manchester University Press, 1956], 11 f.) that exceptions to natural laws do not indicate an imperfection of these laws but God's intention to "supply the . . . wants of grace" are characteristic of the (counterempirical) worldview I have sketched in the text. When uttered it was as unrealistic, from a purely empirical point of view, as the assumption that this world is the best of all possible worlds.

9. Thomas Aquinas, *Summa Theologiae*, Ia,8,4.

10. Ockham's objection is found in his G. de Ockham, *Opera philosophica et theologica*, vol. 1, *Scriptum in Librum Primum Sententiarum ordinatio. Prologus et distinctio prima*, ed. G. Gál and S. E. Brown (Saint Bonaventure, N.Y.: Franciscan Institute, Saint Bonaventure University, 1967), 241. Ockham's epistemology of individuals is directly related to this theological critique.

The result that emerged in the nineteenth century[11] was not a coherent science but a collection of heterogeneous subjects (optics, acoustics, hydrodynamics, elasticity, electricity, heat, etc. in physics; physiology, anatomy, etc. in medicine; morphology, evolutionism, etc. in biology—and so on). Some of the subjects (hydrodynamics, for example) had only a tenuous relation to experiment, others were crudely empirical. It did not matter. Being firmly convinced that the world was uniform and subjected to "inexorable and immutable laws" leading scientists interpreted the collection as an *appearance* concealing a uniform material *reality*.

With the notion of reality I come to the main topic of this essay, which is the relation of human achievements to a world whose features are independent of thought and perception or, to express it more dramatically, the idea that humans are aliens, not natural inhabitants of the universe.

*3. Humans as Aliens*  THE IDEA ASSUMES that the world is divided
*in a Strange World*  into (at least) two regions—a primary region
consisting of important events and a secondary
region that differs from the primary region, blocks our vision, is deceptive
and, in many cases, evil. Grand dichotomies of this kind are found in
many though by no means in all cultures. The dichotomies can be sharpened and made absolute by accidents, social tendencies, abstract reason-

---

11. The situation in the nineteenth century is described in Johann Theodore Merz, *A History of European Thought in the Nineteenth Century,* 4 vols. (New York: Dover, 1965). Christa Jungnickel and Russell McCormmach, *Intellectual Mastery of Nature,* 2 vols. (Chicago: University of Chicago Press, 1986), give a more detailed account of the situation in physics from 1800 to 1925 in German-speaking countries. Daniel Kevles, *The Physicists* (New York: Alfred Knopf, 1978), describes the history of American physics from the second third of the nineteenth century up to the 1970s. See also H. von Helmholtz, "On the Relation of Natural Science to General Science" in his *Popular Lectures on Scientific Subjects,* 1st ser. (New York: Langman Greens, 1898) and E. du Bois-Reymond, "Über die Grenzen des Naturerkennens," in *Zwei Vorträge* (Leipzig: Veit, 1882). Du Bois-Reymond explores the limits of the assumption, declared by Helmholtz to be fundamental for the physical sciences, that explaining the phenomena of Nature means "reducing them to unchangeable attractive or repulsive forces whose intensity depends on the distance" (*Über die Erhaltung der Kraft,* Oswalds Klassiker der exakten Naturwissenschaften, no. 1 [Leipzig: Verlag Wilhelm Engelmann, 1915], 6). Here the gulf between "appearance" and "reality" becomes very noticeable. On the low empirical content of classical hydrodynamics and aerodynamics see L. Prandtl and O. G. Tietjens, *Fundamentals of Hydro- and Aerodynamics* (New York: Dover, 1957), 3. For an opposing point of view see C. Truesdell, *Six Lectures on Modern Natural Philosophy* (Berlin: Springer, 1966), chaps. 1 and 2 as well as chap. 4, p. 85, on Prandtl.

ing, and other agencies. The Gnostic movement, for example, occurred at a time of uncertainty when humans seemed subjected to irrational political and cosmic forces and when help seemed far away. Here are humans "as they really are," i.e., their souls are imprisoned in bodies and the bodies in turn are imprisoned in a material cosmos. This double imprisonment, effected by low-level demons, prevents humans from discovering the truth: the more information they possess about the material world, the more they get involved in it, the less they know. Revelation frees them from their predicament and gives them genuine knowledge.[12]

Realism is part of Genesis. *First* God created the material universe, *then* he created man and woman. The Fall erected a barrier between humans and the world: having been in harmony with nature they are now separated from it. The debates about the nature of Christ (Arianism, Docetism, etc.) deal with the size of the gap.

Greek philosophy is different in some respects, similar in others. Traditional Greek religion distinguished between human and divine properties but without making the distinction absolute: the Greek gods have human features and they participate in human affairs. The distinction hardened when divinities merged as a result of travel and cultural exchange. The gods now lost in individuality but gained in power. Taking his lead from this development Xenophanes introduced a God-monster that does not walk around, does not speak, feels no sympathy but, like a true intellectual, moves the universe by thought alone.[13] Parmenides eliminated the last human property—his principle is pure Being. Using rules familiar from the practice of Near Eastern law he deduced that Being is, that not-Being is not, and that change and difference are illusions.

12. For Gnosticism see Hans Jonas, *The Gnostic Religion* (Boston: Beacon Press, 1958), and R. M. Grant, *Gnosticism and Early Christianity* (New York: Harper and Row, 1966). Early views about the nature of Christ are discussed in Jaroslav Pelikan, *The Christian Tradition*, vol. 1, *The Emergence of the Catholic Tradition* (Chicago: University of Chicago Press, 1971), chaps. 4–6. A more popular account discussing also changes of pictorial representation is Jaroslav Pelikan, *Jesus through the Centuries* (New Haven: Yale University Press, 1985).

13. Xenophanes is discussed in W. K. C. Guthrie, *A History of Greek Philosophy*, vol. 1 (Cambridge: Cambridge University Press, 1962), Parmenides in vol. 2 (1965) of the same work. For Xenophanes see also chap. 2 of P. Feyerabend, *Farewell to Reason* (London: Verso, 1987), for Aristotle chap. 1, sec. 6. In physics the transition to antiparmenidean views is connected with the second law of thermodynamics, the discovery that the most interesting problems in classical mechanics do not lead to integrable systems and chaos theory: Ilya Prigogine, *From Being to Becoming* (San Francisco: Freeman, 1980).

Note that the statement "Being *is*" (*estin* in Parmenides' Greek) was the first conservation law of the West: it asserted the conservation of Being. Like the Gnostics, Parmenides based his results on revelation: it is a goddess that tells him how to proceed.

Now it is interesting to see to what extent the division between two fundamentally different cosmic regions invaded the sciences. Nineteenth-century classical physics—and by this I mean what was then regarded as the fundamental science, classical point mechanics—posited a "real" world without colors, smells, sounds and a minimum of change; all that happens is that certain configurations move reversibly from one moment to another. In a relativistic world even these events are laid out in advance. Here the world

> simply *is*, it does not *happen*. Only to the gaze of my consciousness, crawling upward along the lifeline of my body, does a section of this world come to life as a fleeting image in space which continuously changes in time.[14]

Or, to quote Einstein:

> For us who are convinced physicists, the distinction between past, present, and future has no other meaning than that of an illusion, though a tenacious one.[15]

Note how Einstein's realistic worldview interferes with his empiricism. According to Einstein theories must be tested by experiment. Experiments are temporal processes and, therefore, illusions. But how can illusions inform us about reality? We see here to what extent some modern scientists repeat age-old and, to many, rather disreputable patterns of thought. But remember: modern science with its "inexorable and immutable" laws could not have arisen without these patterns. This is how worldviews can both further and hinder the sciences, depending on circumstances.

*4. Realism as a Worldview and as a Scientific Hypothesis*   THERE IS A WIDESPREAD RUMOR that realism—the idea that the world as laid out in space and time is independent of human perception, thought, and action—has been refuted by delicate but conceptually robust experiments.

---

14. H. Weyl, *Philosophy of Mathematics and Natural Science* (Princeton: Princeton University Press, 1949), 116.

15. Albert Einstein, *Correspondance avec Michele Besso*, ed. P. Speziali (Paris: Hermann, 1979), 312, see also p. 292.

Now if what I have said about worldviews (remember my definition at the end of section 1!) is correct, then the "realism" of the rumor cannot possibly be a worldview. There is no fact, no series of facts, no highly confirmed theory that can dislocate the assumption, made by Einstein, that the events of our lives, experiments included, are nothing but illusions. And even this statement is not adequate. Being tied to individuals and groups a worldview cannot be "Platonized"—it cannot be presented as a person-independent entity that enters into relations with other person-independent entities such as facts and/or theories; it has to be related to the individuals and the communities that are affected by it. And a community holding realism as a worldview simply cannot be shaken by contrary *facts*. If it *is* shaken then this means that it is already breaking up or that the facts presented are part of a powerful rival *worldview*.

Let me illustrate the situation with a historical example. Parmenides' conclusion—there is no change—was contested by his philosophical successors. The situation is usually presented in the following way: Parmenides, positing Being as a single principle, inferred that there is no change. There is change, hence there has to be more than one principle. Democritos assumed infinitely many principles, Aristotle assumed two, potentiality and actuality. The story makes it appear that Parmenides, involved in deep speculation, overlooked change while Democritos, more a man of the world, paid attention to it. But Parmenides was not only familiar with change; in the second part of his poem he even explained some of its features. However, he added that change was not real. To refute him one had to do more than point to changing things. One had to show that change was at least as basic as Being or, to express it differently, one had to show that those addressed by Parmenides possessed a worldview where change played a fundamental role.

This was exactly what Aristotle did. All of us, he said, are citizens—we participate in the actions and deliberations of the city-state; some of us are physicians who try to heal people, others are cobblers, they make boots—our existence is shot through with change and it is this existence that counts, not the results of speculation. Now—can we say that the recent experimental results, which allegedly refute the idea of an objective Nature laid out in space and time, affect scientists and the general public as strongly today as Greek common sense affected Athenians at the time of Aristotle? I don't think so: most scientists are practical realists. At any rate—the matter must be examined; it cannot be taken for granted.

On the other hand we can interpret realism as a hypothesis and not as a worldview. There are passages where Einstein formulates realism in this more modest and more "scientific" manner.

> One cannot simply ask, "Does a definite moment for the decay of a simple atom exist?" but rather instead, "Within the framework of the total theoretical construction, is it reasonable to assume the existence of a definite moment for the decay of a simple atom?"[16]

In other words—can the idea of such a definite moment be incorporated into a theoretical framework that agrees with tests and basic principles? In the passage I have just quoted Einstein does not specify the nature of the tests required. He simply refers to scientific practice. He even rejects, and with some vehemence,[17] the idea that it is possible to compare a theory with a theory-independent "reality": science, not metaphysical speculation, is to decide the question.

In the opinion of most scientists science *has* decided the question—against Einstein's hypothesis. It is somewhat ironical that the decision was prepared by three passionate realists: by Einstein himself, who in his argument of 1935 used an example especially suited for demonstrating the holistic character of quantum mechanics, by Schrödinger, who first noticed this property of the example, and by Bell, whose theorem considerably simplified the demonstration.

Many scientists for whom realism was a worldview and not just a hypothesis now found themselves in a somewhat paradoxical situation. On the one hand they wanted to remain scientists, which means they were inclined to read realism as a hypothesis. On the other hand they had a strong realistic bias. Now science does allow for the continued pursuit of apparently refuted ideas. The survival and eventual triumph of atomism, of the idea that the earth moves, of conservation laws, or, more recently, of matrix mechanics (in the face of well-defined particle tracks) depended on this possibility. But for worldview realists mere possibilities are not good enough. They want stronger support. Trying to obtain it they use phrases that hint at worldviews without bringing them out of the closet. In what may have been his last interview, John Bell said:

16. Albert Einstein, "Reply to Criticism," in *Philosopher-Scientist,* ed. P. A. Schilpp (LaSalle, Ill.: Open Court, 1949), 669.

17. See Arthur Fine, *The Shaky Game* (Chicago: University of Chicago Press, 1986), 93.

For me, it is so reasonable to assume that the photons in those experiments carry with them programs that have been correlated in advance, telling them how to behave. This is so rational that I think, that when Einstein saw that and the others refused to see it, *he* was the rational man. The other people, though history has justified them, were burying their heads in the sand. I feel that Einstein's intellectual superiority over Bohr, in this instance, was enormous, a vast gulf between the man who saw clearly what was needed, and the obscurantist. So, for me it is a pity that Einstein's idea does not work. The reasonable thing just does not work.[18]

"It is so reasonable to assume," says Bell; Einstein "was the rational man," Bohr "the obscurantist." What does that mean? It means that there is a worldview somewhere; that, presented in its full splendor, it would not sound very scientific—and that it is therefore hidden. It is hidden, yes—but from its hiding place it still affects the debate through insinuations, slogans, and attitudes. Most scientific arguments about realism have this truncated character. To complete them we have to reveal the underlying worldview and its relation to the realistic hypothesis. In addition we have to embed the troubling experiments into a rival worldview that is stronger than special professional subjects, gives us a reason to rely on them, and agrees with or even demands the negative outcome of the experiments. Niels Bohr's idea of complementarity contains a sketch of a nonrealistic worldview that satisfies these requirements. Wolfgang Pauli tried to give a more detailed and more complete account.

*5. Wolfgang Pauli's Attempt to Combine Science and Salvation* WOLFGANG PAULI WAS one of the leading physicists of his time.[19] He won the Nobel Prize (for his discovery of the exclusion principle), wrote two surveys which have remained classics (one on relativity, the other on wave mechanics) and technical articles.

18. J. Bernstein, *Quantum Profiles* (Princeton: Princeton University Press, 1991), 84. The Bell-matter is by no means settled. New results may be obtained in moving reference frames (see O. E. Rössler, "Einstein Completion of Quantum Mechanics Made Falsifiable," in W. H. Zurek, ed., *Complexity, Entropy, and the Physics of Information*, SFI Studies in the Sciences of Complexity, vol. 3 (Redwood, Calif.: Addison Wesley, 1990), 367 ff. But even if it were settled there would still be dreams, feelings, religious experiences on the one side and the immovable realism of many scientists on the other.

19. Section 5 contains formulations first used in my column "Atoms and Consciousness" in *Common Knowledge* 1, no. 1 (spring 1992).

Substantial suggestions and criticisms dealing with almost all aspects of the physics of his time are found in his letters, even on postcards. His correspondence had an enormous influence.[20] Some colleagues called him the conscience of physics: others, who seem to have suffered more, its scourge. Pauli delighted in pure knowledge; he despised applications—the "dark backside of science" as he called them—and was highly critical of "the merely rational from whose background a will to power is never entirely absent." He had strong views about world events but refused to participate in collective enterprises.

> My impact should lie in how I *live,* what I *believe* and in the ideas I communicate in a more or less direct way to a small circle of pupils and acquaintances—and not in speaking out before a large public.[21]

Bohr favored political interventions of scientists. Pauli replied:

> Whoever wants to oppose the "Will to Power" in a more spiritual way must not succumb to this Will to such an extent that he ascribes to himself a greater influence on world history than he actually has.

Or, in another letter to Bohr (Pauli's English):

> My attitude is therefore, that we have to be satisfied with the fact—well established by history—that ideas always had a great influence on the course of history and also on politicians, but it is better if we leave the direct actions in politics to other persons and remain on the periphery and not in the center of this dangerous and disagreeable machinery. In [this] attitude . . . I am—last not least—also influenced by the philosophy of Laotse, in which so much emphasis is laid on the indirect action, that his ideal of a good ruler is one whom one does not consciously notice at all.

Pauli wanted to spread his ideas by personal contact with well-prepared friends, not by starting a movement.

This precise individualist was deeply concerned about the direction science had taken since the seventeenth century. To orient himself he examined traditions professional rationalists dismiss with a contemptuous

20. Pauli's observations on the meaningful misuse of physical concepts are contained in "Hintergrundsphysik," in Maier, *Briefwechsel,* 176 ff.

21. Wolfgang Pauli, letter to Max Born (January 21, 1951). This letter and the two that follow are quoted from Ch. P. Enz and K. V. Meyenn, eds., *Wolfgang Pauli, Das Gewissen der Physik* (Braunschweig: Vieweg, 1988), 15 ff.

shrug. He made two points; first, that the rise of modern science was based on a new cosmic feeling and not on experience alone. Kepler, for example, started from the Trinity and ended up with natural laws:

> because he looks at the sun and the planets with the archetypal image in the background he believes with religious fervor in the heliocentric system—*by no means the other way around.*[22]

Pauli's second point was that the development had two branches; one separated humans from a world they were trying to understand and to control, the other sought salvation through a practice (astrology, alchemy, hermeticism) that involved matter and spirit on equal terms.[23] According to Pauli the second branch soon fell apart.

> Today [1955] rationalism has passed its peak. One definitely feels that it is too narrow . . . It seems plausible to abandon the merely rational in whose background the Will to Power always played an important role and to adopt its opposite, for example a Christian or Buddhist mysticism. However I believe that for those who are no longer convinced by a narrow rationalism and whom the spell of a mystical outlook, an outlook that views the external world with its oppressive multitude of events as a mere illusion fails to affect with sufficient strength, I feel that for such people there remains only one choice: to expose themselves fully to these . . . opposites and to their conflict. This is how a scientist can find an inner path to salvation.[24]

Finding a worldview, Pauli seems to say, is a personal matter that must be fought through by every individual; it cannot be settled by "objective" arguments. But only a worldview will enable the individual to make sense of scientific results that seem to run counter to deep-lying (religious, philosophical, scientific) beliefs.

Pauli's correspondence with the Swiss psychiatrist C. G. Jung—which began with his analysis (1931–34, by Erna Rosenbaum, one of Jung's disciples) and lasted until 1957—reveals some stages of this fight.

22. Wolfgang Pauli, "The Influence of Archetypal Ideas on the Scientific Ideas of Kepler," in C. G. Jung and Wolfgang Pauli, *The Interpretation of Nature and the Psyche* (New York: Bollingen Series, 1955), 171.

23. Wolfgang Pauli, *Physik und Erkenntnistheorie* (Braunschweig: Vieweg, 1984), 108 ff.

24. Ibid., 111 f.

It is a rich and complex document, full of surprising and illuminating ideas. Long and detailed descriptions of dreams are used to explore areas incommensurable with natural science but eventually to be united with it. Pauli wrote in 1952:

> More and more it seems to me that the psychophysical problem is the key for the overall intellectual situation of the time; [we can advance by] finding a new ("neutral") unitary psychophysical language for describing an invisible potential reality that can only be guessed at by its effects, *in a symbolic way* [my emphasis].[25]

The key word is "symbolic." As used by Pauli it derives its meaning partly from quantum mechanics, partly from psychology. Quantum mechanics contains terms (the wave function) which seem to refer to things and processes but only serve to arrange phenomena in a systematic way.[26] In his Como Lecture which introduced the idea of complementarity, Bohr called such terms "symbolic." Physical objects are symbolic in an even stronger sense. They present themselves as ingredients of a coherent objective world. For classical physics and the parts of common sense dependent on it this was also their nature. Now, however, they only indicate what happens under particular and precisely restricted circumstances. Combining these two features Pauli envisaged a reality which cannot be directly described but can only be conveyed in an oblique and picturesque way. Making a similar point, Heisenberg wrote:

> Quantum theory is . . . a wonderful example for it shows that one can clearly understand a state of affairs and yet know that one can describe it only in images and similes.[27]

Psychology, long before, had led to analogous conclusions. There are events (apparently senseless actions, dreams, etc.) which, taken by themselves, seem absurd but hint at causes different from their overt appearance. Might it not be possible, asks Pauli, to combine our new physics (matter) and psychology (mind, spirit) by means of characters, namely

25. Wolfgang Pauli, letter to C. G. Jung (1952), in Maier, *Briefwechsel.*
26. Wolfgang Pauli, *Physik und Erkenntnistheorie* (Braunschweig: Vieweg, 1984), 15 n.
27. Werner Heisenberg, *Der Teil und das Ganze* (Munich: R. Piper, 1969), 285.

symbols which play a large role in myth, religion, poetry and thus to heal our fragmented culture?

It is clear that Pauli aimed at a new worldview. The fragmentation he wanted to overcome and the elements of a new unity even announced themselves in his dreams, in an emotionally charged manner. They were both represented by a magical figure, "the stranger." He was young, younger than Pauli. He was a wise man, a magician, conscious of his superiority, even over Pauli, contemptuous of his surroundings, especially of universities, which for him were castles of oppression and which he tried to burn down. He spoke with force and conclusively. Women and children followed him and he began to teach them. Though opposed to science the stranger was tied to its terminology which means that he, too, was in need of salvation. Trying to get his message across he misused scientific terms in a systematic way. Such systematic "misuse," Pauli found, is widespread; it occurs in the writings of engineers, laypeople, and older thinkers. For Pauli this meant that there was a nonphysical reality which, having been robbed of its language (rise of materialism) tried to make itself known in an indirect and "symbolic" manner. Pauli thought that the exploration of this reality would be of importance for the "Western Mind."

*6. What Has Been Achieved?*   TODAY many "Western Minds" are enclosed in bodies that suffer from war, prejudice, illness, hunger, and poverty. Everywhere in the world human beings face problems they cannot solve, not because they lack the right synthesis but because they don't have the power or, after years of suppression, even the will to act. Responding to so much suffering some intellectuals found new ways of making themselves useful. Quantum mechanics and the subtle problems it raises played no role whatsoever in their efforts. Medicine, a new and revolutionary form of Christianity, ecological concerns did; so did the belief that even the most downtrodden individuals know more about reality than was so far credited to them and the corresponding belief that Nature is open to many approaches. We have here a worldview that is not just an intellectual luxury, a kind of delicious dessert one consumes when the trivial matters of daily existence have been dealt with; it is a much-needed response to some of the most pressing problems of our times. I mention this to impress upon you that the relation between quantum mechanics and reality is anything but a universal problem. It is not even a problem for all physicists; it is a problem for a limited group of rather nervous people who assume that their intellectual pains are felt all over the globe.

Wolfgang Pauli's efforts are not therefore irrelevant. True, he did not write for "mankind" (though occasionally he spoke as if he did). Like the field-workers in development, public health, primary environmental care, spiritual guidance, his efforts were for people he knew, his fellow intellectuals who, he thought, might experience problems similar to his own. In a way he even wrote for fighters like Frantz Fanon who was an intellectual and a psychiatrist and who objected to a purely mechanical revival of traditions.

> Such a revival can only give us mummified fragments which because they are static are in fact symbols of negation and outworn contrivances. Culture [a worldview] has never the translucidity of custom [established ideology]; it abhors all simplification. In its essence it is opposed to custom because custom is always the deterioration of culture. The desire to attach oneself to tradition or to bring abandoned traditions to life again does not only mean going against the current of history but also opposing one's own people.[28]

Fanon criticized African intellectuals who were fascinated by Western ways (forms of poetry, for example), who felt guilty, thought they had to do something for their own culture, and started wearing traditional clothes and reviving old customs. But his criticism applies also to other areas. For example, it applies to more recent attempts, in the United States and elsewhere, to replace research by "politically correct" cultural menus, it applies to the rhetoric of a New Age and to those realists who, being faced by the problems of the quantum theory, eschew philosophy but continue to repeat realist slogans. Such actions, says Fanon, do not give us a culture or, as we might say, they do not give us a worldview, something we can live with. They "not only go against the current of history, they also oppose the people one wants to inform." Pauli criticized precisely such tendencies and he tried to overcome them in his letters and in his personal life. True, he looked at the past and he wanted to revive some of its aspects. But connecting the revival with new discoveries and keeping things in flux, he may have contributed to the emergence of a limited, but humane, culture or worldview.

28. Frantz Fanon, *The Wretched of the Earth* (New York: Grove Press, 1963), 224.

# {4} REALISM

*1. Homer and the Bible*   ALMOST EVERYBODY ADMITS that there are dreams, stones, rainbows, fleas, sunrises, murders, planets—and many other things. For the authors of the Homeric epics these entities were real in the sense that they occurred, had distinctive properties, and affected their surroundings. They formed a rich pattern of interactions of varying nature and strength. The effects of dreams, for example, easily surpassed those of trees and stones (the dreams of kings might lead to war and multiple murder). There was no grand dichotomy, such as the dichotomy real/apparent, and events did not conceal or hint at a hidden and perhaps inaccessible world.

Biblical stories, on the other hand, are inscrutable in precisely this sense. Homer, says Erich Auerbach (*Mimesis* [Bern and Munich: Francke Verlag, 1946], 9, 15.) "is all surface . . . nothing is hidden," while a biblical tale is "enigmatic, . . . dark" (17) and "in need of interpretation" (18). There are "many layers, one arranged above the other" (15) and each situation has "subterranean" components that endanger its "clarity" (22).

However, the *Iliad* contains passages where the Homeric world, too, becomes ambiguous and opaque. An example is *Il.* 9.225 ff.

*2. Achilles' Complaint*   HERE ODYSSEUS, acting as an emissary of the Greeks, tries to get Achilles back into the battle against the Trojans. Achilles resists. "Equal fate," he says, "befalls the negligent and the valiant fighter; equal honor goes to the worthless and

Published in C. C. Gould and R. S. Cohen, eds., *Artifacts, Representations, and Social Practice*, 205–22 (Dordrecht: Kluwer Academic Publishers, 1994), © 1994 Kluwer Academic Publishers. Reprinted with kind permission from Kluwer Academic Publishers. Sections 1 and 2 overlap with part I, chapter 1, sections 1–3. Section 3 overlaps with part I, chapter 2, section 5. Section 4 overlaps with part I, chapter 3, section 1. Section 5 overlaps with part II, essay 3.

the virtuous" (318 f.). He seems to say that honor and the appearance of honor are two different things.

Interpreted as a stable, well-defined, and unambiguous entity, the Homeric notion of honor (and the geometric notion that may have corresponded to it) did not allow for such a distinction. Like other epic stereotypes, it was an aggregate containing individual and collective actions and events. Some of the elements of the aggregate were: the role (of the individual possessing or lacking honor) in battle, in the assembly, during internal dissension; his place at public ceremonies; the spoils and gifts he received when the battle was finished; and, naturally, his behavior in all these occasions. Honor was present when (most of) the elements of the aggregate were present, absent otherwise (cf. *Il.* 12.310 ff.—Sarpedon's speech).

Achilles sees things differently. He was offended by Agamemnon, who had taken his gifts. The offense created a conflict between what Achilles received and what he thought was his due. The Greeks who pleaded with Achilles, Odysseus among them, agree that there is a conflict; their suggestions illustrate its customary resolution: the gifts were returned untouched, more gifts were promised, honor is restored (519, 526, 602 f.: "With gifts promised go forth—the Achaeans will honor you as they would an immortal"). So far we are squarely within tradition as expressed in the first part of Sarpedon's speech and reconstructed by our scholars. Achilles is not reassured. Extending the conflict beyond its customary resolution he perceives a lasting clash between honor and its accepted manifestations: honor and the actions that establish and/or acknowledge its presence *always* diverge.

At first sight it seems that we are dealing with a simple case of generalization. Achilles notes a *particular* discrepancy and declares *all* instances of an attribution (or a denial) of honor to be affected by it. There are various difficulties with such an account. To start with, Achilles does not just generalize, he aims at a new *concept;* second, the concept is not a summary of familiar actions and events as are many archaic concepts, it *clashes* with any such summary. Some scholars have added that it also clashes with the *rules of the language* Achilles is using. Achilles, says A. Parry, "has no language to express his disillusionment. Yet he expresses it, and in a remarkable way. He does it by misusing the language he disposes of. He asks questions that cannot be answered and makes demands that cannot be met."[1]

1. A. Parry, "The Language of Achilles," *Transactions and Proceedings of the American Philosophical Association* 87 (1956): 6 ff. The quotation in the next paragraph is from page 7 of the same essay.

But if Achilles "has no language to express his disillusionment," if he "can in no sense, including that of language (unlike, say Hamlet) leave the society which has become alien to him," then what does this so-called disillusionment amount to? We have no description of it, not from Achilles, who, according to Parry, lacks the terms, not from his visitors who are simply confused, and not even from the poet who, after all, provided the evidence for Parry's account. According to Parry and others who believe in the power of linguistic forms (and corresponding forms of life) they are all victims of a "conspiracy" from which there is no escape.[2] Yet Achilles acts as if there were no conspiracy and Parry writes as if he knew the (apparently inexpressible) thought behind Achilles' nonsense. Clearly, there is a mistake somewhere in the argument—but what is it?

Achilles makes statements which sound strange to his visitors and to some twentieth-century commentators. There is nothing unusual about this situation. People often say strange things. But one can ask them and then the matter will either be clarified, or shelved and forgotten until someone else starts asking all over again. Outrage, instant dismissal, lack of interest are other possible reactions.

Nobody can ask Achilles, and his visitors do not pursue the matter. They are puzzled, they argue for a while but they soon give up. This, too, is not unusual. The world is full of garbled messages, unfinished letters, damaged records. Again there are many ways of dealing with the problem, each one having advantages and drawbacks.

The case ceases to be a problem in this familiar, annoying but manageable sense and becomes profound and paradoxical when it is lifted out of its natural habitat and is inserted into a model, a sketch, or a theory of it. A theory (model, sketch) of a historical process says both too little and too much. It says too little because it starts from a fragment of what it wants to represent. But it also says too much because the fragment is not just described, it is subdivided into essence and accident, the essence is generalized and is used to judge the rest. The distinction between linguistic use and misuse and the associated idea of a "conspiracy" both arise

2. "[A]ny language," writes M. Baxandall, *Giotto and the Orators* (Oxford: Clarendon Press, 1971), 44, "not only humanist Latin [the language Baxandall examines in his discussion of how Giotto's work was perceived by the humanists] is a conspiracy against experience in the sense of being a collective attempt to simplify and arrange experience into manageable parcels."

in this way. But this means that the paradoxes I just mentioned—Achilles crosses boundaries of sense as if they did not exist—do not occur in history. They occur when historical events are first "reconstructed" and then evaluated on the basis of unhistorical procedures; they are the mirror image of the ideology and the actions of our scholars.

But although there is no paradox, there is still a conflict: Achilles and his visitors have different ideas about the circumstances that establish or restore honor. The ideas of the visitors are well known and traditional—this, at least, is what the scholars tell us; Achilles' idea is *new*. Not only that; Achilles' new idea seems to explain phenomena which, because of his anger and despair, affect him more than others. The new idea, therefore, has a *structure*, namely, the structure of the phenomena to be explained, "overlaid" by a certain way of relating them to the rest of the world.[3] The structure does not conform to the linguistic habits some scholars, Parry among them, claim to have inferred from the evidence. But stable linguistic habits—if they were indeed stable—are not automatically boundaries of sense—long-lasting empirical beliefs can be and often have been corrected by new discoveries. True, Achilles' visitors are disturbed, they do not agree, perhaps they do not even understand what Achilles is up to. But again we have to ask what this attitude amounts to. Does it show that Achilles talked nonsense, or does it only show that what Achilles said was meaningful, though surprising and unfamiliar? Besides, who is going to judge the matter? Achilles, who does not seem to perceive any difficulty, his visitors who don't say enough to enable us to distinguish between the two possibilities, or a twentieth-century scholar who uses his own parochial distinctions? Considering how much leeway there is, it seems prudent not at once to impose a theoretical frame but to try getting more information from the story itself.

To start with, Achilles' reply does not come out of the blue. It concerns a situation that lies squarely within the common sense of the time—the conflict between custom and Agamemnon's actions. People had violated customs before and had been duly reprimanded. Either such a conflict can be increased or it can be alleviated. And it can be increased either by further actions, e.g., by a further violation, or by seeing the violation in a different light, for example by regarding it as an in-

3. "To exercise a language regularly on some area of experience or activity, however odd one's motives may be, overlays the field after a time with a certain structure." Baxandall, *Giotto and the Orators*, 47.

stance of a general tendency. Achilles, driven by his anger, turns the conflict from a particular disturbance, or a series of disturbances, into a cosmic rift.

I already described the two steps a logician can distinguish in his action. The first step leads from Achilles' personal situation to the situation of many—what can happen to me can and does happen to others. It may sound exaggerated and overly dramatic, but it still conforms to, what I call in a simplifying way, the archaic notion of honor. It is perfectly meaningful. The second step turns the implausible accidental generality so obtained into what one might call an absolute difference. It suggests that honor fails not only because of human weakness, but because of the nature of things: real honor always clashes with its alleged manifestations. We know the motor that makes Achilles speak in this way; it is his disappointment, his anger. What we need to identify are the influences that gave his utterances structure and the analogies, within archaic Greek, that make the structures seem familiar and in this way give meaning to the second step as well.

Homeric thought (and, we may assume, the common sense of the time) was not entirely unprepared for grand subdivisions. Divine knowledge and human knowledge, divine power and human power, human intentions and human speech (an example mentioned by Achilles himself: 312 f.) were opposed to each other in ways that resemble the distinction Achilles wants to introduce. This is the analogy. Achilles uses it; for example, he connects his idea with the judgment of Zeus (607 ff.). But while the judgment of Zeus has a certain arbitrariness and while divine decisions may vary from case to case, the division adumbrated by Achilles has the clarity and power of an objective and person- (God-) independent law. Even this feature has an analogy in the epic: the willfulness of the Gods is never absolute: it is restricted by *moira,* an "irrefrangible order . . . which . . . exists independently of them" and by norms which they may violate but which can be used to criticize and to judge their behavior.[4] Thus Achilles modifies an existing belief by subsuming it under a more general belief that so far had been separated from the

---

4. Walter F. Otto, *The Homeric Gods* (New York: Pantheon, 1954), 276. Cf. also Burkert, *Griechische Religion der archaischen und klassischen Epoche* (Stuttgart: Walter Kohlhammer, 1977), 205 f., and F. M. Cornford, *From Religion to Philosophy* (New York: Harper and Row, 1965), 16. Reference to norms occurs implicitly in the *Iliad,* e.g., in *Il.* 24.33 ff., explicitly in tragedy, for example in Sophocles, *Antigone,* 456.

first. What is the source that not only favors the subsumption (which to later generations seems entirely natural) but maintains the universal validity of the subsuming principle? Or to express it in terms of a later point of view: what prompts Achilles to speak in a way that suggests an as yet unheard-of separation between the social aspect of a property (such as honor) and an independent "real" nature? Do we have to resort to miracles such as "creativity," or are there better ways of explaining the transition?

Had Achilles or the poet who wrote his words lived in the seventh or sixth centuries B.C. I could have answered: the source is closely connected with certain social developments. In politics, abstract groups had replaced neighborhoods (and the concrete relationships they embodied) as the units of political action (Cleisthenes); in economics, money had replaced barter with its attention to context and detail; the relation between military leaders and their soldiers became increasingly impersonal; local Gods merged in the course of travel, tribal and cultural idiosyncrasies were evened out by trade, politics, and other types of international exchange, important parts of life became bland and colorless, and terms tied to specifics accordingly lost in content, or in importance, or they simply disappeared. Without any help from philosophers, politicians, religious leaders, "words . . . bec[a]me impoverished in content, they . . . bec[a]me one-sided . . . empty formulae."[5] One might add that the process was constituted by individual actions, even by conscious decisions which, however, did not have it as its aim and that it was to that extent unconscious. But it modified the "conspiracy" of Homeric Greek by means of a second "conspiracy," actualized some ambiguities of Homeric thought and restructured its content accordingly. Looking at matters "from the outside" we notice that one area of behavior ("Homeric common sense") is being "overlaid" by another (the newly emerging structures just described). Seen "from the inside" we have a discovery: a new important feature of the world is revealed. It is not revealed "as such," but only with respect to the problem that sensitizes Achilles: the existence of personal qualities detached from the efforts of an individual or the reactions of his peers.

But Achilles did not live in the seventh or sixth centuries B.C. He

5. Kurt von Fritz, *Philosophie und sprachlicher Ausdruck bei Demokrit, Platon und Aristoteles* (reprint, Darmstadt: Wissenschaftliche Buchgesellschaft, 1966), 11.

spoke at a time when the developments I enumerated above were in their infancy. They had started; they had not yet produced the results described. Thus Achilles' observations cannot be explained by referring to these results. But they adumbrate them, which means that Achilles' speech also contains an element of invention. It is still discovery, for it reveals the outlines of a slowly rising structure. It is invention because it contributes to the raising structure. It deals with "objective" facts because it is substantiated by a process that is nourished from many sources; it is "subjective" because it is part of the process, not independent confirmation of it. Moreover (this again said in favor of "subjectivity")—the move toward increasing abstraction and the separation of reality and appearance it contained was not the only development that occurred. As becomes clear from funeral inscriptions, passages of comedy, sophistic debates, medical treatises, and especially from the Platonic dialogues,[6] the view that things, ideas, actions, processes are aggregates of (relatively independent) parts and that giving an account of them means giving a detailed but open list remained popular right into the classical age of Greece: "geometric thought" was a seed which grew into many different plants. It is clear that the customary dichotomies (subjective/objective; discovery/invention; etc.) are much too crude to describe complex processes of this kind.

Rather, we have to say that the structures that preceded the "rise of rationalism" were "open" in the sense that they could be modified without being destroyed. They contained the paths Achilles was about to enter, though in a vague and unfinished way. They were also "closed," for it needed a stimulus to reveal ambiguities and alternative structures to reset them. Without the stimulus, words, phrases, rules, patterns of behavior would have seemed clear and unproblematic (clarity is the result of routine, not of special insight); without an (existing, or slowly developing) alternative structure, the possibilities implicit in Achilles' language would have lacked in definition. Thus entities such as "geometric perception" or "the archaic form of life" are to a certain extent chimeras; they seem clear when indulged in without much thought; they dissolve when approached from a new direction. The expression "dissolves," too, is somewhat fictitious—the transition often remains unnoticed and amazes and annoys only a thinker who looks at the process from the safe

6. Details and references in my *Farewell to Reason* (London: Verso, 1987), chap. 3.

distance of a library, or a book-studded office. As always we must be careful not to interpret fault lines in our theories (recent example from physics: the "fault line" that separates classical terms and quantum terms) as fault lines in the world (molecules do not consist of classical parts, and, separated from them, quantum parts). Ambiguity, however, turns out to be an essential companion of change.

But this means that Homer is "clear and superficial" only under special and rather restricted circumstances. His stories become "enigmatic and dark" and even misleading when viewed with a jaundiced eye and approached by way of changing and often unnoticed social tendencies. Was the change effected by Achilles a change for the better? And did Achilles have a choice?

The answer to the second question is simply—no. Achilles participated in a development he was not aware of and could not control. The answer to the first question depends on who is being asked. Achilles himself was not pleased; other participants ran into difficulties. Modern scholars (economic historians, political theoreticians, philosophers) have spoken of progress. Be that as it may—considering the idiosyncratic character of Greek political life there was no way of inhibiting a process that affected all its parts.

*3. Philosophy*     ACHILLES, I SAID, experienced changes he did not create and enlarged dichotomies he had not conceived. Being swept along by events, he lacked the means of opposing the new and, to him, depressing vision of honor and justice. It therefore makes sense to say that his ideas and his actions were "socially conditioned."

The explanation loses much of its power when we turn to the early Greek philosophers. They, too, depended on circumstances they neither noticed nor understood. But these included an element that considerably reduced the impact of ideas, visions and even of the circumstances themselves.

> Far more than their counterparts in most other ancient civilizations, Greek doctors, philosophers, sophists, even mathematicians, were alike faced with an openly competitive situation of great intensity. While the modalities of their rivalries varied, in each the premium—to a greater or lesser degree— was on [novelty], skills of self-justification and self-advertisement and this had far reaching consequences for the way they practiced their investiga-

tions. (G. E. R. Lloyd, *The Revolutions of Wisdom* [Berkeley and Los Angeles: University of California Press, 1987], 98 f.)

Those participating in the competition may have felt that only their ideas deserved attention and that common sense and the notions of their rivals were empty talk. However, to get the attention and to stop the talk they had to consider others, which means that the circumstances of their lives put some distance between their intuitions and their ideas.

"Shepherds of the wilderness," the Muses address Hesiod (*Theogony* 26 ff.), "wretched shameful things, mere bellies, we know how to speak many false things as though they were true; but also know, when we want to, how to utter true things": there is no longer a straight path from impression to truth. Parmenides is guided by a strong vision which occasionally makes him speak with contempt of those who do not share it. Yet he explains, in a series of elementary steps, why his vision should be adopted and regarded as a basic truth. The steps themselves are novel to a certain extent;[7] they are also plausible, which means that truth now depends on the outcome of a battle between plausibilities. Pythagoras made science part of a comprehensive political-religious movement. Yet, having arrived in Kroton he addressed separately the men, the women, and the children of the city and seems to have made a great impression—he knew how to talk to nonphilosophers (Pompeius Trogus in *Iustinus* 20.4). Plato considered various ways of presenting knowledge and, after a detailed analysis of the possibilities of each, used the dialogue, though in a restricted and clearly circumscribed way. Examining a particular topic, Aristotle collected earlier views which then functioned as (modifiable) boundary conditions of his research. The Sophists stayed close to common sense and defended (certain reconstructions of) it against the more abstract systems of their predecessors. "One may assume," write D. E. Gershenson and D. A. Greenberg ("The Physics of the Eleatics," *The Natural Philosopher*, 3 [New York: Blaisdell Publishing Co., 1964], 103, my emphasis),

7. Parmenides' argument employs the form "If . . . then ———." Such forms were familiar from Near Eastern law. The Sumeric code of Ur-Nammu (before 2000 B.C.) used the more complicated clause "if . . . , provided that / / /, then ———" while simple conditionals occur in the Eshnumma Code of 1800 B.C., the Code of Hammurabi, Exodus 21–23, the nonapodictic laws of Deuteronomy and in early Greek law (Gortyn, Draco). Application demanded familiarity with the modus ponens. Parmenides inverted the order of reasoning (indirect proof); this procedure, too, had predecessors.

that the attraction the Eleatic School held for large numbers of its contemporaries and the notoriety it subsequently achieved were due to the reputation its founders had as masters of dialectic in an age which had recently become fascinated by systematic reasoning. *Logic was the rage of the day;* all over, in the marketplaces, in the streets, in private homes and in public buildings, at all times, sometimes all through the night, people engaged in dialectical disputations and flocked to hear the acknowledged masters of logical argument display their art.

What fascinated the Athenians—for Athens was now the stage on which the different approaches collided—was the multiplicity of the ideas proposed, the strange nature of some of them, and the possibility of proving a thesis as well as its opposite (cf. Plato, *Euthydemus* 257d ff.). There were many offerings; a person in search of knowledge had to choose not only among results but among methods of argumentation as well. I shall illustrate the problem by discussing some aspects of Parmenides' philosophy.

*4. Parmenides, Atomism, Common Sense, and Medicine*   THE IONIAN COSMOLOGISTS postulated a unity behind diversity and tried to reduce observable events to it. They began with plausible physical hypotheses. Thales, for example, identified the unity with water. He may have argued for his choice; for example, he may have said that water occurs in a solid, a fluid, and an airy form and that it is necessary for life. And we must not forget that the Greeks, who lived around the Mediterranean "like frogs around a pond," had firsthand experience of the immensity and the many functions of water (mist, air, clouds, etc.). Thales' successors suggested other substances, again with plausible reasons. Parmenides pointed out that the physical approach was rather shortsighted. According to Parmenides the most basic entity underlying everything there is, including any hypothetical substance one might propose, is Being. Speaking with hindsight, we can say that this was a shrewd suggestion: Being is the place where logic and existence meet. Every statement involving the word "is" is also a statement about the essence of the world.

What can we say about Being? That it is *(estin)* and that not-Being is not. What happens on the basic level? Nothing; the only alternative to Being is not-Being, not-Being does not exist, hence there is no change. How is the basic level structured? It is full, continuous, without subdivisions. Any subdivision would be between Being and something else, the only something else on the basic level is again not-Being, not-Being does not exist, hence there are no subdivisions.

But is it not true that we traditionally assume and personally experience change and difference? Yes, we do. Which shows, according to Parmenides, that neither tradition (*ethos poly'peiron*—"habit, born of much experience," Diels-Kranz 87, fragment 3) nor experience ("the aimless eye, the echoing ear . . .," 87, 5) is a reliable guide to knowledge. This was the first, the clearest and most radical separation of domains which later were called reality and appearance and, with it, the first and most radical defense of a realist position. It was also the first theory of knowledge.

Those who are ready to make fun of Parmenides should consider that large parts of modern science are bowdlerized versions of his result. His premise, estin—Being is—is the first explicit conservation law; it states the conservation of Being. Used in the form that nothing comes from nothing (which found its way into poetry: *King Lear* 1.1.90) or, in Latin, *ex nihilo ni[hi]l fit*, it suggested more specific conservation laws such as the conservation of matter (Lavoisier) and the conservation of energy (R. Mayer).[8] The more philosophically inclined practitioners of nineteenth-century physics posited a "real" world without colors, smells, etc., and with a minimum of change; all that happens is that certain configurations move reversibly from one moment to another. In a relativistic world even these events are laid out in advance. Here the world

> simply *is*, it does not *happen*. Only to the gaze of my consciousness, crawling upward along the lifeline of my body, does a section of this world come to life as a fleeting image in space which continuously changes in time. (H. Weyl, *Philosophy of Mathematics and Natural Science* [Princeton: Princeton University Press, 1949], 116)

"For us, who are convinced physicists," wrote Einstein

> the distinction between past, present, and future has no other meaning than that of an illusion, though a tenacious one. (*Correspondance avec Michèle Besso*, ed. P. Speziali [Paris: Hermann, 1979], 312; cf. also 292)

Irreversibility, accordingly, was ascribed to the observer, not to nature herself. And so on. None of the scientists who supported the dichotomy

8. "Die organische Bewegung in ihrem Zusammenhang mit dem Stoffwechsel," in R. Mayer, *Die Mechanik der Wärme*, Oswalds Klassiker der exakten Wissenschaften, no. 180 (Leipzig: Verlag Wilhelm Engelmann, 1911), 9.

could offer arguments that were as simple, clear, and compelling as those of Parmenides.

And yet Leucippus, who is traditionally associated either with the Eleatics in general or with Parmenides' disciple Zeno,

> thought he had a theory which harmonized with sense perception and would not abolish either coming-to-be and passing-away or motion and the multiplicity of things. He made these concessions to the facts of perception. On the other hand, he conceded to the Monists that there could be no motion without a void. The result is a theory which he states as follows: "The void is a 'not-Being,' and no part of 'what is' is a 'not-Being'; for what 'is' in the strict sense of the term is an absolute *plenum.* This *plenum*, however, is not 'one'; on the contrary, it is a 'many,' infinite in number and invisible owing to the minuteness of their bulk . . ." (Aristotle, *De generatione et corruptione* 325a23 ff., Ross translation)

In short: Being is many and moves in not-Being. Note the nature of the argument: Leucippus does not try to refute Parmenides by using the "fact" of motion. Parmenides had been aware of the "fact" and had declared it to be illusory. Moreover, he had not simply asserted the illusory character of motion, he had presented proofs. He had transcended sense impression on the grounds "that 'one ought to follow the argument'" (Aristotle, *De generatione et corruptione* 325a12 f.). Leucippus, in contrast, decided to follow perception; one might say that he and those who thought in a similar manner (Democrites, Empedocles, Anaxagoras) wanted to bring physics closer to common sense.

A general principle that supports the move was formulated by Aristotle. Commenting on the Platonists who justified the virtues by reference to a supreme Good, he wrote (*Ethica Nicomachea* 1096b33 ff., my emphasis):

> Even if there existed a Good that is one and can be predicated generally or that exists separately and in and for itself, it would be clear that such a Good can neither be produced nor acquired by human beings. *However, it is just such a Good that we are looking for . . .* One cannot see what use a weaver or a carpenter will have for his own profession from knowing the Good itself or how somebody will become a better physician or a better general once "he has had a look at the idea of the Good" [apparently an ironical quotation of a formula much used in the Platonic school]. It seems that the physician does not try to find health in itself, but the health of human beings or perhaps even the health of an individual person. For he heals the individual.

Thus we can say that at the time in question (fifth to fourth century B.C.) there existed at least three different ways of establishing what is real: one could "follow the argument"; one could "follow experience"; and one could choose what played an important role in the kind of life one wanted to lead. Correspondingly there existed *three notions of reality* which differed not so much because research had as yet failed to eliminate falsehoods but because there were different ideas as to what constituted research.

Following his argument Parmenides established a reality that was "objective" in the sense that it was untouched by human idiosyncrasy. Following his different approach, Aristotle introduced a reality that depended on the nature, on the achievements, and, especially, on the *interests* of humans. Leucippus, Democritus, and others had an intermediate position; they moved toward common sense but stopped early on the way. Still, their results clashed with established subjects such as medicine.

Empedocles, for example, used four elements, the Hot, the Cold, the Moist, the Dry. Change is defined as a mixing and unmixing of the elements, health as their balance, illness as their imbalance. Considering these definitions the author of *Ancient Medicine,* who seems to have been a practicing physician, wrote as follows (chapter 15):

> I am at a loss to understand how those who maintain the other [more theoretical] view and abandon the old method [of direct inspection] in order to rest the *techne* on a postulate [i.e., who introduce abstract principles such as the elements of Empedocles] treat their patients on the lines of this postulate. For they have not discovered, I think, an absolute cold and hot, dry and moist that participates in no other form. On the contrary, they have at their disposal the same foods and the same drinks we all use, and to the one they add the attribute of being hot, to another, cold, to another, dry, to another, moist, since it would be futile to order patients to take something hot, as he would at once ask "what hot thing?" So they must either talk nonsense [i.e., speak in terms of their theories], or have recourse to one of the known substances [i.e., add their descriptions in an ad hoc manner to common practice].[9]

Aristotle himself produced "intermediate" (see above) subjects and made them a measure of existence. *"We physicists,"* he wrote (*Physica* 185a12 ff.)

9. The text is known as one of the so-called Hippocratic writings and can be found in *De prisca medicina,* chap. 15 (*Corpus Medicorum Graecorum* 1.1, p. 46 line 18–p. 47 line 11).

must take for granted that the things that exist by nature are either some or all of them in motion . . . Besides, no man of science is bound to solve all the difficulties that may be raised but only [those that arise from the mistaken use of] the principles of [his] science.

Adding the "intermediate" notions we obtain not just three conceptions of reality, but an indefinite number. It seems that the "problem of reality" has many solutions.

*5. Modern Science*     "THAT IS QUITE UNDERSTANDABLE," the modern reader will reply. "What you are describing is a period before the rise of modern science. But modern science is (1) based on a uniform approach, has (2) led to a coherent body of results which (3) force us to make science not just *a* measure, but *the* measure of reality." Neither (1) nor (2) nor (3) is correct.

As I have argued elsewhere,[10] scientists from different areas use different procedures and construct their theories in different ways; in other words—they, too, have different conceptions of reality. However, they not only speculate; they also test their conceptions and they often succeed: the different conceptions of reality that occur in the sciences have empirical backing. *This is a historical fact, not a philosophical position* and it can be supported by a closer look at scientific practice. Here we find scientists (Luria in molecular biology, Heber Curtis, Victor Ambarzumian, Halton Arp, and Margaret Geller in astrophysics and cosmology, L. Prandtl in hydrodynamics, etc.) who want to tie research to events permitting "strong inferences," "predictions that will be strongly supported and sharply rejected by a clear-cut experimental step" (S. E. Luria, *A Slot Machine, a Broken Test Tube* [New York: Harper and Row, 1985, 115) and who show a considerable "lack of enthusiasm in the 'big problems' of the Universe or of the early earth or in the concentration of carbon dioxide in the upper atmosphere," all subjects that are "loaded with weak inferences" (Luria, 119). In a way these scientists are continuing the Aristotelian approach, which demands close contact with experience and objects rather than following a plausible idea to the bitter end.

However, this was precisely the procedure adopted by Einstein (Brownian motion, general relativity); by the researchers in celestial me-

10. See the essay "Has the Scientific View of the World a Special Status Compared with Other Views?" (part II, essay 2 of this volume). *Ed*

chanics between Newton and Poincaré (stability of the planetary system); by the proponents of the atomic theory in antiquity and later, down to the nineteenth century; by Heisenberg during the initial stages of matrix mechanics (when it seemed to clash with the existence of well-defined particle tracks); and by almost all cosmologists. "Is it not strange," asks Einstein (letter to Max Born, in *The Born-Einstein Letters* [London: Macmillan, 1971], 192),

> that human beings are normally deaf to the strongest argument while they are always inclined to overestimate measuring accuracies?

—but just such an "overestimating of measuring accuracies" is the rule in epidemiology, demography, genetics, spectroscopy, and other subjects.

I repeat that all the subjects I just mentioned have been successful, thus confirming the notions of reality implicit in their theories. Even outlandish conjectures that ran counter to physical common sense were confirmed. An early example is Maxwell's calculation of the viscosity of gases. For Maxwell this was an exercise in theoretical mechanics, an extension of his work on the rings of Saturn. Neither he nor his contemporaries believed the result—that viscosity remains constant over a wide range of density—and there was contrary evidence. Yet more precise measurements turned the apparent failure into a striking success. It pays to "follow the argument."[11]

This is true even for large-scale enterprises such as the entire history of modern science. Surrounded by comets, new stars, plagues, strange geological shapes, unknown illnesses that would come out of the blue, linger for a while, and disappear again, comets which behaved in exactly the same manner, irrational wars, biological malformations, meteors, oddities of weather, the leaders of Western science asserted the "inexorable and immutable" character of the basic laws of nature (Galileo to Castelli, letter of December 21, 1613). Early Chinese thinkers had taken the empirical variety at face value. They had favored diversification and had collected anomalies instead of trying to explain them away. Aristotelians had emphasized the local character of regularities and insisted on a classification by multiple essences and corresponding accidents. Sci-

---

11. Maxwell's calculations are reproduced in *The Scientific Papers of James Clerk Maxwell*, ed. E. D. Niven (1890; reprint, New York: Dover Publications, 1965), 377 ff. For more recent examples cf. G. Birkhoff, *Hydrodynamics* (New York: Dover Publications, 1955), secs. 20 and 21.

entists of an equally empirical bent, Tycho Brahe among them, regarded some anomalous events of their time as miracles or, as we might say to-day, they took cosmic idiosyncrasies seriously; others, like Kepler, as-cribed them to the subjective reactions of the telluric soul while the great Newton, for empirical as well as for theological reasons, saw in them the finger of God (*Opticks,* query 31).

Empiricism and early versions of instrumentalism were supported by leading theologians. Saint Thomas (*Summa Theologica,* question 8, art. 4) had assumed that "inexorable and immutable" universals existed side by side with God, only in a different manner. Duns Scotus and Ockham objected that all we have are the results of God's acts of will. We do not understand why these acts occurred, we do not know what acts will oc-cur in the future, we can only observe their effects, connect them, and hope for the best.[12] It needed a tremendous act of faith to continue "fol-lowing the argument" and not only to *assume* regularities where no reg-ularities had yet been found, but *to work toward them* for centuries, and in the face of numerous failures.

The result that emerged in the nineteenth century was not a coherent science but a collection of heterogeneous subjects (optics, acoustics, hy-drodynamics, elasticity, electricity, heat, etc., in physics; physiology, anatomy, and practical knowledge, in medicine; morphology, evolution-ism, etc., in biology—and so on). Some of the subjects (hydrodynamics, for example) had only the most tenuous relation to experiment, others were crudely empirical. There were surprising unifications such as Max-well's unification of magnetism, optics, and electricity, but they soon created problems (problem of the reference system) while the allegedly most basic subject of all, point mechanics, ran into conceptual difficul-ties of the most serious kind (cf. Boltzmann's introduction to his *Vor-lesungen über die Principe der Mechanik* [Leipzig: Johann Ambrosius Barth, 1897]). Johann Theodore Merz, *A History of European Thought in the Nineteenth Century* (1904–1912, reprint, New York: Dover Publica-tions, 1976) has described the situation in a precise and detailed manner.

In the twentieth century the overall situation is roughly the same. One must not be misled by the fact that scientists, when asked, produce a more or less uniform response. For the ways in which they think and pro-

12. G. de Ockham, *Opera philosophica et theologica,* vol. 1, *Scriptum in Librum Primum Sententiarum ordinatio. Prologus et distinctio prima,* ed. G. Gál and S. E. Brown (Saint Bonaventure, N.Y.: Franciscan Institute, Saint Bonaventure University, 1967), 241.

ceed in their own domains (elasticity, ethology, the chemistry of substances, rheology, linguistics, etc., etc.—elementary particle physics is not the only science!) are as varied as they were before. Thus all we can offer today is a *list,* similar to the list given by Merz, enumerating the achievements and the drawbacks of various approaches, and we can identify "science" with this complex and somewhat scattered war on many fronts. Alternatively we can put one view on top and *subordinate* the others to it, either by pseudo-derivations (most so-called "reductions" dissolve when one looks at them more carefully), or by declaring them to be meaningless. We can also present a *paste job* where each particular view and the results it has achieved are smoothly connected with the rest thus producing the impression of a single and coherent scientific "reality." I prefer to think that

> [t]here is no simple "scientific" map of reality—or if there were, it would be much too complicated and unwieldy to be grasped or used by anyone. But there are many different maps of reality, from a variety of scientific viewpoints. (John Ziman, *Teaching and Learning about Science and Society* [Cambridge: Cambridge University Press, 1980], 19)

or, putting reality where the achievements are, there are different kinds of reality defined by different modes of successful research which in turn are backed, more or less consciously, by different ideas about the relation between humans and their surroundings: assumptions (1) and (2) are false. As regards (3), it suffices to point out that practical results, taken *by themselves,* never forced anyone to accept one view of reality over another. The artisans of ancient Greece, who extracted metals from the ore, prepared them for various purposes, manufactured weapons, designed jewelry, built houses, bridges, ships, theaters, made important contributions to ancient culture. They had detailed information about materials of the most varied kind. Yet the philosophers did not start from this information, but from some very abstract principles.[13] In book 7 of his *Republic,* Plato enumerates the subjects that might be useful for the future leaders of the state: music, arithmetic, geometry, astronomy. Music is useful because it furthers harmony and grace, arithmetic aids the general in distributing his troops, geometry helps him in demarcating a military camp

---

13. Similarly the historians used legends, not local historical traditions, as the starting points of their research. Cf. Felix Jacoby, *Griechische Historiker* (Stuttgart: Alfred Druckmueller Verlag, 1956), 220, right column, lines 64 ff.

and the barracks within, while astronomy is needed for orientation and the calendar. Plato adds that the first three subjects have also a theoretical side where numbers, lines, and sounds are related not to material things, but to each other. According to Plato the resulting structures form an unchanging real world whose perception creates knowledge and stabilizes the mind.

This distinction between *abstract theory* which provides understanding and *practical subjects* which are never entirely transparent played an important role in the rise of modern science. It replaced the rich *practical information about materials* by *theories of matter* which were abstract and poor in content,[14] but defined reality and guided research for generations to come: practical results, taken by themselves, do not enforce any particular notion of reality. Note, incidentally, that abstract theory triumphed in the end, which shows that an undeflected preference for practical knowledge would have been most impractical.

*6. A Task for Philosophers*

NOW IF SCIENCE is indeed a collection of different approaches, some successful, others wildly speculative, then there is no reason why I should disregard what happens outside of it. Many traditions and cultures, some of them wildly "unscientific" (they address divinities, consult oracles, conduct "meaningless" rites to improve mind and body) succeed in the sense that they enable their members to live a moderately rich and fulfilling life. Using this extended criterion of success I conclude that non-scientific notions, too, receive a response from Nature, that Nature is more complex than a belief in the uniformity and unique excellence of science would suggest, and that an interesting task for a writer capable of looking beyond the limits of a particular school would be to consider some of its properties. I myself have started from what Pseudo-Dionysius Areopagita said about the names of God. God, he said, is in-

14. Physicians and other early opponents of the excesses of philosophy (cf. the above quotation from *Ancient Medicine*) expressed their objections in writing—they were members of a tradition of written exchange that was soon to dominate Western civilization. Not all crafts participated in this tradition; we have no written reports from potters, metal workers, architects, miners, painters, etc. We must reconstruct their knowledge from their work and from indirect references to it. Cyril Stanley Smith, a metallurgist from MIT, did this in a book (*A Search for Structure* [Cambridge: MIT Press, 1981]) as well as in an exhibition (photographic displays in *From Art to Science* [Cambridge: MIT Press, 1980]).

effable. But depending on our approach God may respond in a variety of comprehensible ways. God is not identical with any one of these ways and it would be a mistake to identify Him (Her, It) with, say, Nature as conceived by modern cosmology (there are also personal manifestations of divinity). Moreover, describing a response and not Being itself, all knowledge about the world now becomes ambiguous and transparent. It points beyond itself to other types of knowledge and, together with them, to an unknown and forever unknowable Basic Reality. Thus the literary forms used by the composers of the Bible seem far better adapted to our situation than the more lucid but basically superficial stories that have replaced them.

# {5}  HISTORICAL COMMENTS

## ON REALISM

THE DEVELOPMENTS caused by Bell's inequality have further weakened the case of realism. Bell himself, who never concealed his realist preferences,[1] admitted that "history has justified them [the orthodox]. Einstein was the rational man. The other people . . . were burying their heads in the sand. I feel that Einstein's intellectual superiority over Bohr's in this instance [the EPR correlations] was enormous; a vast gulf between the man who saw clearly what was needed and the obscurantist."[2] "So, for me, it is a pity that Einstein's idea does not work. The reasonable thing just does not work."[3] Confronted with this new difficulty the realists repeat what they have said before:[4] that realism, on general grounds, is preferable to idealism and positivism, that there exist promising realistic models, and, especially, that relativity is not an obstacle. Few scientists are impressed. The majority prefers clear theorems and decisive experiments to promises and declarations of faith. Both parties *make prophecies*, the orthodox that observer dependence has come to stay, the realists that their models will eventually produce em-

Published in A. van der Merwe, F. Selleri, and G. Tarozzi, eds., *Bell's Theorem and the Foundations of Modern Physics* (Singapore: World Scientific, 1992), 194–202. Reprinted with kind permission. The first three pages of this essay overlap with part I, chapter 3, section 1, and with part II, essay 4, section 1. There is also some overlap with part II, essay 7.

1. J. S. Bell, *Speakable and Unspeakable in Quantum Mechanics* (Cambridge: Cambridge University Press, 1987), 41, 52, 126, and 170; "J. Bell," in P. C. W. Davies and J. R. Brown, eds., *The Ghost in the Atom* (Cambridge: Cambridge University Press, 1986), 45–57; Jeremy Bernstein, *Quantum Profiles* (Princeton: Princeton University Press, 1991), chap. 1.

2. "Indeed, I have very little idea what this means," writes Bell, commenting on Bohr's reply to EPR; *Speakable and Unspeakable* 155.

3. Bernstein, *Quantum Profiles*, 84.

4. Survey in Franco Selleri, *Die Debatte über die Quantentheorie* (Braunschweig: Vieweg, 1990); and "J. Bell" (45–57) and "David Bohm" (118–34), in Davies and Brown, *The Ghost in the Atom.*

pirical results. But the study of prophecies belongs to history, not to physics. Let us therefore see what history can contribute to the debate.

The separation of subject and object or, more generally, of appearance and reality arose (in the West), between 900 and 600 B.C. as part of a general movement toward abstractness and monotony. Money replaced gift giving and an exchange of goods, local gods merged, gained in power but lost in concreteness and humanity, abstract laws, not family relations, defined the role of citizens in a democracy, wars were increasingly fought by professional soldiers—and so on. Language changed accordingly. The rich vocabularies that had described the relation between humans and their surroundings shrunk, some terms disappeared, others converged in meaning. All this just occurred, without any explicit and clearly planned contribution from individuals and special groups. The new habits, the older and more idiosyncratic ways of doing things, and the features implied by both were all equally *real*—they were not dreams or apparitions. However, they were not equally *important*. Special groups, soon to be called philosophers, turned importance and universality into measures of existence—in the following way.

The Ionian cosmologists had postulated a unity behind diversity and had tried to reduce observable events to it. They began with plausible physical hypotheses. Thales, for example, identified the unity with water. He may have argued for his choice, for example, he may have said that water occurs in a solid, a fluid, and an airy form and that it is necessary for life. And we must not forget that the Greeks, who lived around the Mediterranean "like frogs around a pond," had firsthand experience of the immensity and the many functions of water. Thales' successors preferred other substances such as fire, or air. Then Parmenides pointed out that the physical approach was rather shortsighted. The most basic entity underlying everything there is, including any hypothetical substance one might propose, is Being. This was a very shrewd suggestion, for Being is a place where logic and existence meet: Every logical statement involving the word "is" is also a statement about the essence of the world. What can we say about Being? That it is *(estin)* and that not-Being is not. What happens on the basic level? Nothing. The only possible change of Being is into not-Being, not-Being does not exist, hence there is no change. How is the basic level structured? It is full, continuous, without subdivisions. Any subdivision would be between Being and something else, the only something else on the basic level is not-Being.

Not-Being does not exist, hence there are no subdivisions. But is it not true that we assume and experience change and difference? Yes, we do. Which shows, according to Parmenides, that neither tradition nor experience is a reliable guide to knowledge. This was the first, the clearest and the most radical separation of domains which later were called reality and appearance. It was also the first theory of knowledge.

Those who are ready to make fun of Parmenides should consider that large parts of modern science are bowdlerized versions of his result. To start with, his premise, *estin*—Being is—is the first explicit conservation law; it states the conservation of Being. Used in the form that nothing comes from nothing (which found its way into poetry: *King Lear* 1.1.90) or, in Latin, *ex nihilo nihil fit*, it suggested more specific conservation laws such as the conservation of matter (Lavoisier), and the conservation of energy (R. Mayer, who started a decisive paper with this very principle). Nineteenth century classical physics posited a "real" world without colors, smells, etc., and a minimum of change; all that happens is that certain configurations move reversibly from one moment to another. In a relativistic world even these events are laid out in advance. Here the world "simply *is*, it does not *happen.* Only to the gaze of my consciousness, crawling upward along the lifeline of my body, does a section of this world come to life as a fleeting image in space which continuously changes in time."[5] "For us," wrote Einstein, "who are convinced physicists, the distinction between past, present and future has no other meaning than that of an illusion, though a tenacious one." Irreversibility, accordingly, is ascribed to the observer, not to nature itself. And so on. None of the scientists who support a dichotomy of this kind can offer arguments which are as simple, clear, and compelling as those of Parmenides and nobody has explained how deceptive appearances and "illusions" can inform us about a real world that excludes them. Even worse—the "immediate sense impressions"[6] of Planck, Einstein, and

5. H. Weyl, *Philosophy of Mathematics and Natural Science* (Princeton: Princeton University Press, 1949), 116. The Einstein quotation is from *Albert Einstein, Correspondance avec Michele Besso,* ed. P. Speziali (Paris: Hermann, 1979), 312; cf. also 292. For the resistance against basic change, cf. I. Prigogine, *From Being to Becoming* (San Francisco: Freeman, 1980).

6. A. Einstein, letter to M. Solovine of May 7, 1952, quoted from A. P. French, ed., *Einstein: A Centenary Volume* (Cambridge: Cambridge University Press, 1979), 270. According to Einstein ("Physics and Reality," quoted in *Ideas and Opinions by Albert Einstein* [New York; Crown Publishers, 1954], 291 f., written in the years of the Einstein-

other empiricists are not part of our experience (which is an experience of objects in space), but have to be unearthed by special methods (reduction screen, etc.). Thus we have here a view in which a hidden reality thoroughly independent of human events is said to be based on hidden processes extremely dependent on them. One cannot say that things have improved since Parmenides.

They have not improved, practically oriented scientists reply, because Planck and Einstein were enmeshed in philosophical speculation. Evidence can be identified, objective processes discerned without a general account of either. Definitions such as the famous definition of an "element of physical reality" of EPR belong to a special scientific debate, they make sense within that debate, but it would be foolish trying to nail them down forever.

This judgment, which makes scientific practice, not philosophical speculation, the measure of method, evidence, and reality, has also philosophical predecessors, the much-despised Aristotle among them. Aristotle criticized Parmenides in two ways. He tried to show the mistakes in Parmenides' reasoning and he pointed out that change, which Parmenides had called unreal, is important in human life. "Even if there existed a Good that is one and can be predicated generally, or that exists separately and in and for itself," Aristotle wrote about similar ideas in Plato (*Ethica Nicomachea* 1096b33 ff—my emphasis), "it would be clear that such a Good can neither be produced nor acquired by human Beings. *However, it is just such a Good we are looking for* . . . one cannot see what use a weaver or carpenter will have for his own profession from knowing the Good in itself or how somebody will become a better physician or a better general once 'he has had a look at the idea of the Good'

---

Podolsky-Rosen correlations), we start from a "labyrinth of sense impressions," select from it "mentally and arbitrarily, certain repeatedly occurring complexes of sense impressions," correlate them to the concept of a bodily object, and attribute to this concept "a significance, which is to a high degree independent of the sense impressions which originally give rise to it": reality is a "mental and arbitrary" construction, introduced "to orient ourselves" in a "labyrinth of sense impressions." Now, first of all, such a "labyrinth" is nowhere to be found in our lives; it is itself an "arbitrary" construction. Second, how can we introduce the order we need "to orient ourselves" while still lacking it, i.e., while being disoriented? It is not surprising that Max Planck, who defended a similar version of realism, called it "irrational" and "somewhat contradictory": "Positivismus and reale Aussenwelt," first read in 1930 and quoted from *Vorträge und Erinnerungen* (Darmstadt: Wissenschaftliche Buchgesellschaft, 1969), 234.

[apparently an ironical quotation of a formula much used in the Platonic school]. It seems that the physician does not try to find health in itself, but the health of human beings or perhaps even the health of an individual. For he heals the individual."

Assume, says Aristotle, that somebody had succeeded in proving that there are unchanging entities ("even if there existed a Good," he says). Assume furthermore that our lives, though stable to some extent, don't show a trace of them. Then we can either try to adapt to the entities, or continue on a way that has the strength of experience and tradition behind it. The first alternative is chosen by religious sects. Taking particular and often quite limited events as their guide, the members of such sects reorient their entire existence. An example I already mentioned are the early philosophers; they conferred reality on ideas which not only went beyond experience but also clashed with it. Scientists who believed in "inexorable and immutable laws" (Galileo), when the evidence showed an abundance of exceptions, did the same. Their sect had important ancestors (Platonists, Stoics, and Cartesians), and it gradually changed into an empirical enterprise. Modern science could not have arisen without it.

According to Aristotle, the citizens of Athens made a different choice. They not merely accepted change, they maintained it by their actions. As I pointed out, there was social power behind the tendency toward abstraction that culminated in Parmenides. But this power was overruled by the power of change and diversity. Using the word "real" to describe what is basic for an individual, a group, a nation, we can say that change, for the Athenians, was very real indeed or, generalizing: *real is what plays a central role in the kind of life we identify with.* I shall now develop some consequence of this principle, or of *Aristotle's principle* as I shall call it.[7]

7. A recent defender of Aristotle's principle is Schrödinger. Criticizing Bohr's point of view, he wrote (letter to W. Wien of August 25, 1926, published in W. Wien, *Aus dem Leben eines Physikers* [Leipzig: Barth, 1930], 74, and quoted from V. V. Raman and Paul Forman, "Why Was It Schrödinger Who Developed de Broglie's Ideas?" *Historical Studies in the Physical Sciences,* vol. 1 [Philadelphia: University of Pennsylvania Press, 1969], 301): "Physics consists not merely of atomic research, science not merely of physics, and life not merely of science. The purpose of atomic research is to fit our experiences from this field into the rest of our thought; but all the rest of our thought insofar as it has to do with the external world, moves in space and time." Democritus used the principle even before Aristotle. "Having declared his distrust in sense perception by saying: 'In our

A *first* and rather immediate consequence is that the boundary between reality and appearance cannot be established by scientific research; it contains a normative or, if you will, an "existential" component.

This explains, *second,* why so many different processes (visions, immediate experience, dreams, and religious fantasies) have been declared to be real and why discussions about reality produce so much heat. After all, they are debates about the right way to live or, in more narrow domains, about the right way of doing research. They deal with the weight to be given to reason, experience, emotion, faith, fascination, and further entities which in some views are strictly separated while they merge in others.

*Third,* different ways of life entail different interpretations of expert knowledge or, more recently, of scientific knowledge. Theologians like Saint Thomas and philosophers like Descartes and Leibnitz regarded natural laws as the work of a stable and reliable divine being, of a genuine rationalist. Statements expressing such laws were therefore objective and necessarily true. Duns Scotus and William of Ockham, both critics of Saint Thomas, emphasized the immense power and the unfathomable will of God, which manifest themselves in individual events. One can observe these events, one can summarize the observations in general statements, but one cannot go further. Natural laws, accordingly, are about observations and about nothing else. Which view is correct? That depends. If the world, whether divine or material, is as described by Ockham, then there are no objective laws and instrumentalism is correct. But is it not the task of science to decide the question and to establish one interpretation to the exclusion of all others?

It is not, because, *fourth,* science contains different traditions (atomism and more phenomenological approaches are examples from the past) and, besides, it is not the only source of knowledge. People arranging their existence around nonscientific phenomena and declaring them to be real did not end in disaster—at least not all of them did. They developed detailed and effective cultures. Applying Aristotle's principle to each and every one of these cultures, we arrive at a form of relativism:

usual ways of talking we say that there is colour, sweetness, bitterness, in truth, however there are only atoms and the void,' he lets the senses reply to reason: 'poor reason; from us you took your evidence and want to knock us down with it? Our loss will be your fail'"
(Galen, *De medicis empiricis fragmentum,* quoted in Demokrit, in Diels-Kranz, *Die Fragmente der Vorsokratiker* [Zurich: Weidmann, 1985], fragment 68 B125).

there is more than one way of living and, therefore, more than one type of reality. However, while traditional relativists infer truth and reality from the *mere existence* of criteria, perceptions, procedures, beliefs, Aristotle's principle invites us to add *success* and to explain it by assuming a deeper lying stratum that responds positively to many different endeavors.

It follows, *fifth*, that the sciences are incomplete and fragmentary. One sees this in a more direct way when considering the large areas of experience and human action that constitute the lives of past and present generations but are regarded as unscientific, subjective, and irrational. In these circumstances it makes no sense to look for "the" correct interpretation of, say, quantum mechanics. And, indeed, there exists a great variety of interpretations, corresponding to different worldviews.

Taken literally, quantum mechanics comes close to Parmenides: There are no well-defined objects and no distinguishable observations which can be stored. Movements that find salvation in overcoming change and difference welcome this feature and want to preserve it (first interpretation). Bohr, more a man of the world, rejected the consequence. The experimentalist, he said, must be separated from his surroundings and described in commonsense or, if precision is needed, in classical terms. Humans are "detached observers." Like Aristotle Bohr made a choice and interpreted quantum mechanics accordingly (second interpretation).

But we don't experience ourselves, our actions and our relation to the world in a detached way. Bohr himself often emphasized that we are both actors and observers on the stage of life. Wolfgang Pauli, who took this aspect very seriously, tried to bring physics closer to it. "According to my point of view," he wrote (letter to Bohr of February 15, 1955) "the degree of [the] 'detachment' [of the observer] is gradually lessened in our theoretical explanation of nature and I am expecting further steps in this direction." Physical objects are not simply there; they "assume an ambiguous . . . and symbolic character" while the wave function is a "symbol that unites mutually incompatible perceptual contents" (third interpretation).[8] And then, of course, we can still imagine underlying "objective" mechanisms (fourth interpretation).

8. Quoted from K. V. Laurikainen, *Beyond the Atom* (New York: Springer, 1988), 60, and W. Pauli, *Physik and Erkenntnistheorie* (Braunschweig: Vieweg, 1984), 98 (detached observer), 15 (symbolic nature of objects), and 15 n. (symbolic nature of wave function).

Each interpretation adapts empirical and mathematical results to a wider perspective and a corresponding conception of reality. If reality is identified with perceptible physical events which are sharply separated from the perceiving subjects, then a coherent account of reality is impossible and Bohr is correct. If, on the other hand, we regard reality as hidden and coherent and its manifestations as fragments, then interpretations such as those of Bohm (explication of an implicit order) will sound eminently reasonable. Bohm's views do not clash with facts—they clash with a certain view about the role of facts, namely, that facts are parts, not manifestations of what there is. According to Pauli, the objects of quantum mechanics are too tied to special circumstances—the very rigid conditions of individual experimentation and large-scale projects—to permit an inference about all there is (there are also fear, pity, and the unconscious, and nobody knows if these can be cut off from matter without repercussions). They are not elementary building stones of the world. But they can serve as hints, or analogies.

Pauli's views have much in common with the general picture that emerged from Aristotle's principle. In this picture we start with a world (which I shall call the primal world, or Being) which behaves in its own way and not necessarily in accordance with any one of the laws that have been discovered by scientists. (Here we still have an element of realism.) Humans are part of the primal world, not detached aliens, and they are subjected to its whims: Being can send scientists on a wild-goose chase—for centuries. On the other hand, it permits partial independence (cf. item five above) and it provides some of those acting independently (not all of them!) with *manifest worlds* they can expand, explore, and survive in (manifest worlds are in many respects like ecological niches). Inhabitants of a particular manifest world often identify it with Being. They thereby turn local problems into cosmic disasters. But the manifest worlds themselves demonstrate their fragmentary character; they harbor events which should not be there and which are classified away with some embarrassment (example: the separation of the arts and

---

Cf. also the discussion at the symposium "The Copenhagen Interpretation 60 Years after the Como Lecture" (August 6–8, 1987, Joensuu, Finland), report published by the Department of Physical Sciences, University of Turku, Finland, March 1988, especially the second section. Cf. also C. A. Maier, ed., *Wolfgang Pauli und C. G. Jung: Ein Briefwechsel 1932–1958* (Berlin: Springer, 1992). Max Planck admitted that his reality was "detached" from the world of observations: "Positivismus," 235.

the sciences). The transition from one manifest world to another cannot be described in either except by excising large regions originally thought to be real—a good case for applying the notion of complementarity.[9] Bell's request that a fundamental theory should not contain any reference to observation is satisfied, but trivially so. Being as it is, independently of any kind of approach, can never be known, which means that really fundamental theories don't exist.

9. Cf. Niels Bohr, "Natural Philosophy and Human Cultures," *Nature* 143, no. 268 (1939).

# { 6 }  W H A T   R E A L I T Y ?

I SHALL TRY TO APPROACH what some people call the issue of reality in a series of steps, using examples as I go along. The first step deals with the ways in which the word "reality" might be used.

*1. Jason and Medea*     HAVING ARRIVED IN CORINTH and having married the daughter of Creon the king, Jason confronts Medea. She had fallen in love with him, had helped him to obtain the Golden Fleece, and had killed her brother to protect his escape. Feeling abandoned and alone in a strange country where she is despised as a barbarian and feared as a sorceress, this proud princess had threatened the king and had been exiled. Jason reproaches her for her unreasonable behavior. He calls himself her friend and promises that, like a responsible person, he will take care of her and her children. In a long speech (520–75) he explains his views and justifies his actions. It was not she who helped him, but Cypris (one of the names of Aphrodite) who watched over his adventures and protected his life. Not freely choosing but driven by the power of love, Medea had used her great intelligence—and she is being amply rewarded for it. Instead of living among the barbarians she can now live in Greece where the law, not crude force, guides human action. Her talents are well known—but nobody would have mentioned them had she remained at the ends of the earth. Besides—his marriage to the princess was a clever move; it brought security not only to him, but also to her and her children, allowing them "the most important thing of all," to live in honor and without want.

First published in Italian as "Quale realtà?" in Antonello La Vergata and Alessandro Pagnini, eds., *Storia della filosofia, storia della scienza: Saggi in onore di Paolo Rossi* (Florence: La Nuova Italia, 1995), 79–91. The article "Potentially Every Culture Is All Cultures" (published in *Common Knowledge* 3, no. 2 [fall 1994]) is related to section 6 of the present essay.

Were I to produce the play I would present Jason as a considerate, patient, though somewhat conceited gentleman—a former Homeric hero, not averse to seeking out danger and using others, now retired, a fugitive, because of the machinations of his uncle and because of Medea's way of dealing with them, but still shrewd enough to gain the protection of a king and the love of his daughter. Jason listens to Medea, he tries to understand her, he does not abandon her but takes what he thinks are her needs into consideration and kindly, but with increasing exasperation, explains the matter point for point. However—and here lies the great art of Euripides as understood today—the kinder and the more rational he behaves the more cruel and cowardly he appears to Medea. And this is not just a matter of opinions. It is her form of life that makes Medea perceive Jason in this manner.

For Jason and Medea have two different and subtly articulated worldviews, the worldviews clash, and disaster is the result. Or, describing realities rather than views about it (cf. Aristotle, *De poetica*, chap. 9) we have two ways of living, acting, perceiving, and understanding—the heroic way of life and a woman's view (objectivized by the chorus)—and they clash. Conflicts of this kind had been described before, for example in Aeschylus's *Oresteia*. Here the clash between traditional laws and the new law of Zeus and Apollo leads to a paradox: there exist actions which imply impossible results whether or not they are carried out—an early and rather interesting application of a reductio ad absurdum. The paradox is removed by the divinely supervised vote of an assembly of Athenian citizens, i.e., by consulting opinions. But after that the power of Athena enforces the New Order, lifting it from the domain of opinions into the domain of objective social constraints or, as one might say, of reality. One can of course restrict reality to material processes. That would make important events very unreal indeed.

*2. Tables and Chairs*     YET JUST SUCH A RESTRICTION is either demanded or implied by philosophers who regard physics, biology, and other "materialistic" sciences as the basis of everything.

Thus trying to become clear about the notion of reality Arthur Fine appeals to common sense:

> I certainly trust the evidence of my senses as a whole with regard to the existence and features of everyday objects. And I have similar confidence in the system of "check, double-check, triple-check" of scientific investigation. ( *The Shaky Game* [Chicago: University of Chicago Press, 1986], 126 f.)

In other words: real in science is what is linked to experiment by steps analogous to those by which everyday objects are linked to "the evidence of the senses." But what are everyday objects and what features are we talking about?

Academic champions of common sense hardly ever get beyond tables and chairs (this includes Fine: 128 n. 18). And even here they consider only the most tiresome properties: a table is *brown*, it has a *solid* top, and we can infer that it has a *backside*. But a furniture designer, an art historian, or a couple about to furnish their apartment are interested in very different matters: how do the chairs and the tables fit into the new surroundings? Does the table cling to the floor thus conveying stability or is it ready to take off like a nervous bird? Are the chairs friendly? Are they looking forward to receiving the bulging hind ends of visitors or are they hostile to being sat on? What is more important—the aesthetic impression of a chair, its connection with what an artist thinks are the tendencies of the times (cf. Charles Eames, *Chaise*—never commercially produced), or its function as a sitting machine? Don't say the properties I just mentioned are subjective and therefore irrelevant. Their presence or absence can make or break a furniture designer, it can create or destroy the atmosphere of a home. Besides, one can learn to perceive these properties just as one can learn to perceive *Daphnia pulex* amidst the chaos one finds when first looking into a microscope. It is true that the general attitude toward the properties changes—this is why there are styles, fashions and why they have a history. But during the rule of a particular style the properties themselves are fairly stable or can be stabilized by perceptual training. After that they join the "features of everyday objects" and can be established by the "evidence of the senses as a whole"—no matter what some people may think about their ontological status. One might as well say that they are real properties, in a rather straightforward sense of the word "real."

Two comments. First, the sense I am referring to was used by Aristotle in his criticism of Parmenides' philosophy, which, translated into modern terminology, implies that reality is one, unchanging, and indivisible and that change and subdivision are unreal. On the contrary, said Aristotle (see e.g. *Physica* 185a12 f., *Ethica Nicomachea* 1096b33 ff., *De generatione et corruptione* 325a18 ff.), "it would be next door to lunacy" to regard as unreal what plays such an important role in our lives. Second, tables and chairs are not as important as large buildings and especially churches. The impact of these objects was—and to some extent still is—much more compelling, not only for individuals but for entire traditions and in that sense even more real.[1]

*3. Faces*  BUT TABLES AND CHAIRS are not the most important everyday objects—people are and, more especially, their attitudes as expressed in their gait and in their faces. Let us therefore take a closer look at faces.

In his *Analyse der Empfindungen* (Jena: Fisher, 1906, 3 n. 1) Ernst Mach describes the following phenomenon:

> As a young man I once saw in the street a face in profile which I found highly disturbing and repulsive. I was shocked when I discovered that it was my own face which I had perceived by way of two mutually inclined mirrors. On a later occasion I was rather tired after a strenuous nocturnal journey on a train. Entering a bus I saw another person entering from the opposite side. "What a dilapidated schoolmaster!" I thought. Again it was I, for I had faced a large mirror.

How shall we interpret this phenomenon? Shall we say that, being unprejudiced, the first impression gives us the real character of Ernst Mach? Or shall we prefer the second impression, which is the result of a lifetime of observations?

Some time ago I visited a friend of mine whom I had not seen for about fifteen years. I rang the bell, the door opened, and I faced a plump, smallish, gray-haired woman. "She got herself a housekeeper," I thought—but only for a moment, for I soon recognized her. Immediately the phenomenon grew, the face became younger and looked almost as it had looked fifteen years earlier. Which impression reflected "reality"? The first impression, which was the impression of an unknown woman, but correctly placed on a qualitative age scale, or the second impression, which was wrong, agewise, but contained the many experiences we had shared? What about my dreams, where she turned up with a different face entirely but clearly recognizable? What about photographs, paintings, icons which are about the same person but show different and conflicting features? Formerly blind people can identify their friends and their pets in the chaos of sensations they perceive when first being able

---

1. For some of the ideas involved cf. Otto von Simson, *The Gothic Cathedral* (New York: Bollingen Foundation, 1962).

to see; what they notice are almost shapeless blobs—but they clearly represent the persons they know from nonvisual personal contact. Considering this plethora of phenomena and insisting on a precise relation between evidence and "reality," one is forced to conclude that "the evidence of my senses as a whole with regard to" faces is incoherent not only in itself, but also with the evidence of others, and that the "reality" inferred may be a chimera. Poets (Pirandello, for example) have described this situation in an impressive way.

But family members, the inhabitants of a village, the members of closely knit professional groups (of actors, scientists, business associates, soldiers working behind enemy lines) do come to a decision about the character of those with whom they live and base their actions and, occasionally, their lives on what they have decided. Obviously, they do not "proceed scientifically" in the sense of forming hypotheses related to a clearly defined body of evidence. Can psychological tests restore scientific order? They may—but that is not the point. For the question is if individuals, villagers, the members of closely knit groups would be better off using, say, the Minnesota Multiphasic Personal Inventory instead of what they are accustomed to. This we do not know and shall never know—there is no way, short of tyranny, of obtaining the fair samples that would be needed for a scientific examination. Besides, even a strict scientific test with suitably chosen samples would depend on what is to be regarded as a "better life." Some people might prefer frequent quarrels to a graveyard peace.

I conclude that there are large areas where the question of what is real and what is not (and, therefore, of what is true and what is not) not only lacks an answer but cannot be answered from the nature of the case. Those who believe in a uniform world and who do not want to break the connection with experience must therefore regard the phenomena I described as confused appearances of a reality that can never be known.

In the next section I shall examine some implications of the above idea. Before that, however, I would like to emphasize that *in practice* questions of reality are decided in a very different way. What usually happens is that powerful groups treat ideas that belong to their research program as if they were pieces of reality. If they have influence, then, their reality may become a basis of educational, political, medical reform. This is how religion kept people captive and this is how views about intelligence, human nature, mind, and body were used to keep society clean.[2] Just recently Daniel Koshland, the editor of *Science,* was asked if the large amounts of money spent on the human genome project should not better be used to relieve the misery of the homeless. His reply: "What these people [i.e., those who asked the question] don't realize is that the homeless are impaired," meaning that homelessness was a genetic, not a political or cultural, problem and should be dealt with by molecular-biological methods. It does not matter that this assertion goes far beyond what is known in molecular biology.[3] PR got the project going; further PR provides a steady flow of money to those engaged in it.

*4. Reality as Ineffable*   IT WAS A SIGN of enlightened tolerance (drama-tized with great skill in Lessing's *Nathan der Weise*) to declare that though Judaism, Christianity, and Islam differed in many ways, they yet revered the same Divine Being. But such an attitude could only be upheld by disregarding important details. Marcion and the Gnostics had noted the glaring differences between the God of the Old and the New Testaments and had postulated (at least) two Gods, only one of them benevolent. A monotheist who rejects vagueness, metaphor, and analogy is forced to admit that though we may have accounts of how God *appeared* to different groups, we have no description and no knowledge of *God Himself.* This move can be readily extended to the notion of a uniform reality.

Many cultures assume that social events occur in surroundings which may be out of reach for humans but which still affect their lives, even to the extent of providing the material and the forms that constitute a human being. For some cultures the surroundings are close, epistemologically speaking, and can be explored by just looking around. For others they are concealed by deceptive appearances and accessible to special methods only. (Parmenides, the Gnostics, and leading scientists belong to the second group.) Considering the role they play in our lives one might call such surroundings *reality* and separate them from the (in the second case, often erroneous) opinions and appearances about them.

2. Stephen Jay Gould has described some aspects of this process in his *The Mismea-sure of Man* (Harmonsworth: Penguin, 1984).

3. Cf. R. C. Lewontin, "The Dream of the Human Genome," *New York Review of Books,* May 28, 1992, with references.

As in religion there exist different views about the nature of this reality. As in religion this abundance is often removed by fiat: one view is correct, the rest is deception. And as in religion there exist arguments for such a judgment. The arguments hardly affect the faithful—their beliefs have an entirely different foundation. But they serve to appease unruly intellectuals who occasionally have the power to make or break a movement.

In the sciences we have mainly two arguments for the privileged position of scientific views: they are "rational," and they are successful.

The first difficulty with these arguments is that science is a mixed bag of opinions, procedures, "facts," "principles," not a coherent unit. Different subjects (anthropology, psychology, biology, hydrodynamics, cosmology, etc., etc.) and different schools within the same subject (empirical and theoretical trends in astrophysics, cosmology, and hydrodynamics; phenomenology and "high theory" in elementary particle physics; morphology, embryology, molecular biology, etc., in biology—and so on) use widely differing procedures, have different worldviews, argue about them—*and have results:* nature seems to respond positively to many approaches, not only to one.[4] Considering this plethora of ideologies, approaches, facts some philosophers and sociologists no longer worry about how to separate, say, physics from biology but wonder how the whole business stays together, not only administratively, but also conceptually.[5]

We can go further. Many cultures were (and are) successful in the sense that they guaranteed the mental and physical well-being of their members. This was known at times when cultures still learned from each other (as, for example, the ancient Greeks learned from their neighbors in the Near East) and it is gradually being rediscovered by anthropologists, development workers, and historians of agriculture, medicine, architecture, astronomy who marvel at the many ways in which "even primitives" could survive and flourish in places which to Western eyes are barren and unsuitable for human life. Facing the negative consequences of the expansion of Western civilizations into areas that had

---

4. For details and references see "Has the Scientific View of the World a Special Status as Compared with Other Views?", essay 2 in part II of the present volume.

5. Example: the essays in A. Pickering, ed., *Science as Practice and Culture* (Chicago: University of Chicago Press, 1992).

been independent and self-supporting scientists developed methods and approaches which use the knowledge and the skills of local communities to improve the situation. An enormous literature explains the problems and describes the results.[6] The outcome is that the argument from success (which, incidentally, is ready to admit that success is never complete and that cultures can always profit from each other) selects not only "science" but other procedures as well, including the worldviews and cosmologies that have developed in their wake: nonscientific worldviews are as good candidates for having grasped reality as is science.

The matter of rationality, however, is easily dealt with. Rationality played an important role in building up the scientific approach, which then moved away from it. This means that "rationalism" (which by now is as complex and scattered as "science") is neither independent nor acceptable evidence for the sciences. They have to stand on their own feet.

Now if what I have said so far is correct, i.e., if not only one but many different stories about our surroundings have to be taken seriously, then Lessing's problem of the three rings arises with renewed force. For unitarians the solution is the same as above—we have evidence *how Being reacts* when approached in different ways, but *Being itself and the conditions of its acting* in a certain way remain forever shrouded in darkness.

*5. Outside Support*     THIS SOLUTION can be connected with and supported by a variety of points of view. One is the point of view that emerged from quantum mechanics: properties once believed to be "in the world" depend on the approach chosen, and the instrument connecting the results of the various approaches, the wave function has only a "symbolic" function (Bohr in his Como lecture). Physical objects are symbolic in an even stronger sense. They appear as ingredients of a coherent, objective world. For classical physics and the parts of common sense associated with it this was also their nature. Now, however, they only indicate what happens under particular and precisely restricted circumstances. Combining these two features Wolfgang Pauli envisaged a reality that cannot be directly described but can only be con-

6. For a lively survey with references consult John Reader, *Man on Earth* (Harmondsworth: Penguin Books, 1988); for a very interesting special comparison between "primitive" and scientific assertions cf. Gerard Gill, "But How Does It Compare with the Real Data?" *RRA Notes* 14 (1992): 5–14.

veyed in an oblique and picturesque way. "Quantum theory," writes Heisenberg on this matter (*Der Teil und das Ganze* [Munich: R. Piper and Co. Verlag, 1969], 285)

> is . . . a wonderful example of the situation that one can clearly understand a state of affairs and yet know that one can describe it only in images and similes.

Psychology, long before, had arrived at a similar conclusion. There are events (apparently senseless actions, dreams, etc.) which, taken by themselves, seem absurd but hint at causes different from their overt appearance. Freud's dream symbolism was a first attempt to clarify at least part of the situation. Might it not be possible, Pauli asked in a variety of articles and in his correspondence, just published, with C. G. Jung, to combine our new physics (matter) and psychology (mind, spirit), by means of characters, namely symbols, which also play a role in myth, religion, and poetry, and thus heal our fragmented culture?

The second point of view that fits the situation I am trying to explain is the philosophy of Pseudo-Dionysius Areopagita, an otherwise unknown author who wrote about A.D. 500 and had a tremendous influence on theology, politics, and architecture: the cathedral of Saint Denis, which anticipated the Gothic style, was built with his ideas in mind.[7] According to Pseudo-Dionysius, God (or, using the terms of this paper, Ultimate Reality, or Being) is ineffable. Concentrating our entire strength on Ultimate Reality we face nothingness, a void, no positive response (Ultimate Being, says Hegel, "ist in der Tat *Nichts,* und nicht mehr noch weniger als Nichts").[8] But we can describe and explain our interaction with certain emanations of God or, to express it in a less theological manner, we have access to the ways in which Ultimate Reality reacts to our approach. Ultimate Reality, if such an entity can be postulated, is ineffable. What we do know are the various forms of *manifest reality,* i.e., the complex ways in which Ultimate Reality acts in the domain (the "ontological niche") of human life. Many scientists identify the particular manifest reality they have developed with Ultimate Reality. This is simply a mistake.

7. Details in E. Panofsky, *Abbot Suger on the Abbey Church of St. Denis and Its Art Treasures* (Princeton: Princeton University Press, 1946).

8. "[I]ndeed is *nothing,* and not more nor less than nothing": *Logik,* pt. 1, chap. 1, B.

*6. Potentially Every* THE IDEA THAT REALITY is uniform but ineffa-
*Culture Is All Cultures* ble is not the only possible way of bringing order into what we think we know. Another way
which, as far as I am concerned, is less one-sided (though compatible
with the uniformity thesis) would be to admit that there are many different kinds of objects and features, that they are related to each other in
complex ways, that some of them, such as fashions in architecture, furniture, and dress, reflect human interests while others, though manufactured with the help of complex equipment, seem to be more independent, and that this hierarchy becomes the more obscure the more we try
to remove ourselves from it. So far a unitarian realism claiming to possess positive knowledge about Ultimate Reality has succeeded only by
excluding large areas of phenomena or by declaring, without proof, that
they could be reduced to basic theory, which, in this connection, means
elementary particle physics. An ontological (epistemological) pluralism
seems closer to the facts and to human nature.

I just spoke of an "ontological pluralism"; like most people I, too, am
liable to summarize complex stories by using simple, though learned-
looking, terms. I therefore have no right to complain when others import the term "relativism" and call me a relativist. But I can still correct
them, in the following manner.

To start with, not all approaches to "reality" are successful. Like unfit
mutations, some approaches linger for a while—their agents suffer,
many die—and then disappear. Thus the mere existence of a society with
certain ways of behaving and certain criteria of judging what has been
achieved is not sufficient for establishing a manifest reality; what is also
needed is that God, or Being, or Basic Reality reacts in a positive way.[9]
Whatever relativism seems to occur in this paper is therefore not a philosophical position; it is an empirical fact supported by the multiplicity of
approaches and results within and outside the sciences.

Second, traditional relativism assumes that cultures are "closed" and
well defined—they can deal with every question by either answering it or
by regarding it as nonsensical. But this is not how "real" cultures react.

9. This caveat, incidentally, is very old. Already Protagoras insisted that truth,
though defined by what appears to an individual, is not always connected with happiness
(see Plato, *Theaetetus* 166d1 ff.).

Facing sizeable problems (or long-lasting successes) *they change;* the society that solves a major problem is not the society that ran into it and therefore cannot count as a stable measure of success. In a way we can say that *potentially every culture is all cultures.* Moreover, the successes I have been describing and the cultures to which I have attributed them are temporary events, never well defined, always ambiguous. Just remember that "science" in the sense in which it is understood today arose in the nineteenth century, that the separation of theory and experimental practice occurred toward the end of that century, that mammoth enterprises such as the experiments carried out in CERN or in the Abruzzi are just a few decades old, that they have changed the very foundation of science (objectivity no longer means repeatability) and that science may change again, as a result of superstrings, chaos theory, and similar inventions.

# {7}  ARISTOTLE

ARISTOTLE WAS A SCIENTIST, a philosopher, and a historian. He
engaged in research, defined the nature of science, and showed
how it could be adapted to the needs of public life. He founded
and advanced a variety of subjects from basic physics via biology, psychology, politics, sociology, economics and the humanities to poetic theory. He realized that knowledge may affect character and has to be used
with care. "Any occupation, art, science," he wrote, "which makes the
body, or soul, or mind less fit for the practice or exercise of virtue, is vulgar . . . some [kinds of knowledge] are proper for a free man but only to
a certain extent, and if he attends to them too closely, in order to attain
perfection in them, the same evil effect will follow": a humane science
must be adapted to the requirements of a balanced and rewarding life.

According to Aristotle these requirements also define its content. Aristotle lived at a time when subjects such as medicine, rhetoric, astronomy—even cooking—which had proven themselves in practice were
criticized for lacking a theoretical foundation and when new and highly
abstract notions of knowledge came to the fore. Like their modern successors the ancient theoreticians dismissed personal experience and
ridiculed "mere" fact collectors. Aristotle applauded their drive toward
unity but denied that the unities achieved were more real than the facts
unified. Commenting on Platonists who had tried to explain, to justify,
and, perhaps, to correct the behavior of artisans and common citizens by
reference to a supreme Good, he wrote (my emphasis):

> Even if there existed a Good that is one and can be predicated generally or
> that exists separately and in and for itself, it would be clear that such a Good
> can be neither produced nor acquired by human beings. *However, it is just*

Undated essay; previous publication history unknown. This essay has some overlap
with part I, chapter 3, section 2.

*such a Good that we are looking for* . . . one cannot see what use a weaver or a carpenter will have for his own profession from knowing the Good in itself or how somebody will become a better physician or a better general once "he has had a look at the idea of the Good" [apparently an ironical quotation of a formula much used in the Platonic school]. It seems that the physician does not try to find health in itself, but the health of human beings or perhaps even the health of an individual person. For he heals the individual.

In other words: a Universal Good should reflect the reality of the individual benefits that are collected under its name, not the other way around. Commenting on theoreticians such as Parmenides or the atomists, who denied (certain kinds of) change on the grounds that their own basic theoretical entities did not change, Aristotle pointed out that "natural things," i.e., the things that shape our lives, "are some or all of them subject to change": a particular mode of existence, the waking state of a healthy and responsible human being is made the measure of truth and reality.

This is a most interesting procedure. Aristotle neither examines the argument of the theoreticians (he did that, too, but in a different context) nor does he confront it with some theorizing of his own. *He rejects the whole approach.* The task of thought, he seems to say, is to comprehend and perhaps to improve what we perceive and do when engaged in our ordinary everyday affairs; it is not to wander off into a no-man's-land of abstract and empirically inaccessible concepts.

The view that there are different levels of being has a long history; it arose independently of philosophy and the sciences. There are religions and myths which distinguish between an ordinary kind of existence and a higher world. Ordinary existence is always with us: it is the way we live; special procedures (fasting, mental exercises, rituals) and divine help are needed to escape from it and to enter the higher world. The early Western philosophers, Parmenides among them, retained the view, introduced a new kind of ritual—argument—but still conceded the need for divine guidance (the divine element has survived until today, though in a secularized version; it is now called "creativity"). Parmenides' argument was simple and convincing: the most basic ingredient of the world is *Being;* the only property of Being is *that it is;* the only change which Being can undergo is to cease to be, i.e., to turn into not-Being; but not-Being does not exist; therefore, change does not exist and experience which suggests that it does is not a reliable guide to knowledge. Note that Parmenides' property is an early and rather sweeping conservation principle: Being always remains the same, it always *is*. Later and more restricted conservation principles were supported by analogous considerations: Robert Mayer, a discoverer of energy conservation, started one of his essays with the blunt statement "nothing comes from nothing" and then tried to identify the something that remains unchanged. The energeticists of the next generation, Wilhelm Ostwald among them, tried to explain physical processes from the energy principle alone and for a long time denied the need for separate change-producing principles. Einstein's theory of relativity has Parmenidean features: there is a basic reality, the four-dimensional space-time continuum with all world lines laid out in advance. Change occurs when a subject moves along one of the world lines—it is mere appearance. The idea of universal and unchanging laws of nature which can be found in Descartes and Leibnitz and which reached a high point late in the nineteenth century has similar features. We are not too far from the truth when saying that modern debates about the relation between science and common sense are but the most recent version of an age-old tension between religions that sanctify the here and now and other religions where fulfillment occurs (knowledge is obtained) in distant regions. Aristotle's procedure resolves the debates in favor of (a streamlined form of) common sense. It accepts the simplifications and unifications achieved by scientists but denies that they point toward an uncommon reality. Those who postulate such a reality, an Aristotelian would say, make an assumption not forced upon them by the evidence. (For example, the validity of the energy principle does not commit us to the view that there exists a *substance*, energy, in addition to the constancy of the transformation ratios.) It is true that scientists often interpret their theories in a naïvely realistic way and that their realism has led to amazing discoveries. But what aids a researcher in solving recondite problems need not be beneficial for society at large (soldiers in battle may be encouraged by the idea that life is a contest and that the only thing that counts is winning—but the same idea would be disastrous when used as a foundation for a civilized society). Besides, interpretations that led to fruitful theories were often discarded later on while the theories survived (Schrödinger's interpretation of his wave mechanics is a case in point). We have to conclude that reality cannot be left to the specialists. But whom should we ask instead?

According to Aristotle the use and the interpretation of scientific results are a political matter. Now, politics concerns the life of citizens as expressed in their varying experiences and their common sense. Hence, reality is measured by this experience and common sense. The measure need not be stable: the citizens may find a myth, or a story claiming scientific credentials so attractive that they decide to see the world in its terms. The early Christians are one example. The acceptance of scientific beliefs by "educated" citizens of the sixteenth and seventeenth centuries is another. Aristotle's analysis makes it clear that in the latter case we are dealing with political decisions superimposed upon scientific arguments. An overriding respect for experts tends to blur the distinction between the political (or, when the decision is automatic, the social) and the scientific elements of our notions of reality: we are inclined to believe that the pronouncements of the experts are knowledge of the purest kind, without any admixture. A study of Aristotle, aided perhaps by a study of Duhem's exposition, restores clarity and returns to the citizens a power they relinquished by mistake.

Aristotle was for a long time the victim of uninformed gossip. Even specialists who ought to have known better ascribed to him views he never held and criticized him for crimes he never committed. This started early in the thirteenth century, when Aristotle's major writings on natural philosophy returned to European intellectual life. They were condemned, and lists of Aristotelian propositions regarded as heretical were compiled. Aristotle survived the theologians of Paris; he barely survived the rise of modern science. In the nineteenth century he was studied by classical scholars, by metaphysicians, but only rarely by scientists. Many twentieth-century writers regard him as a symbol of dogmatism and backwardness. "No modern scientist would consult Aristotle!" is the battle cry of scientists and philosophers of science blinded by the belief in the incredible novelty of modern scientific thought. But the situation is changing. Historians have shown that Aristotle's scientific ideas were used and bore fruit long after the triumph of Copernicanism. Scientists, realizing the barrenness of a crudely reductionistic attitude, look favorably upon the holistic features of Aristotle's physics (which is a general theory of change, including generation and corruption, qualitative change such as the transfer of information from a wise teacher to an ignorant pupil—an example frequently used by Aristotle himself—and locomotion). Historians of biology and biologists praise the extraordinary intuition and conceptual power Aristotle brought to the study of living organisms. Some of them prefer Aristotle to modern mathematical approaches such as catastrophe theory. Aristotle's interpretation of the continuum as a whole whose parts are created by cuts (temporary halts in the case of motion) and cannot be said to exist before a cut has been performed implies that a well-defined location and a well-defined state of motion exclude each other; this anticipates some very profound features of modern physics. Galileo rejected Aristotle's account as making no difference to the length of a line: a line will have the same length whether it is now a whole, or an assembly of points. The reply is that though it may have the same length it does not have the same structure and that the difference reveals itself in problems such as Zeno's paradoxes and the problems of blackbody radiation, first resolved by Planck.

Finally, I would like to show how Aristotle's reputation was damaged by superficial arguments and a fairy-tale history. It is a commonplace, repeated in many textbooks, that Aristotle, using speculation, denied a vacuum and that he was refuted by Toricelli's and Guericke's experiments. Now, first of all, it is very difficult to see how it can ever be shown, "by experiment," that an observed volume has nothing in it. There will at least be light, as Leibnitz pointed out, in his debate, via Clarke, with Newton. A brief look at Guericke's procedure shows the extent of the difficulty. Guericke starts with a promise: he will put an end to the fruitless debates between the plenists and the defenders of a vacuum. Experiment will resolve the issue. Experiment showed that no volume can be completely voided of matter. There always remains a residue. Guericke ascribes the residue to matter that evaporates from the bodies of the experiment. The residue will be the smaller, the farther away the bodies, hence, he concludes, there must be a vacuum somewhere in interstellar space. Note, first, the highly speculative procedure. Experiments are promised, they are carried out, but their results are of the desired nature only if we extrapolate a special and unexamined hypothesis about them (evaporative tendencies of matter) into interstellar space. Note, second, that the procedure proves a vacuum only if matter can already be assumed to consist of atoms separated by a vacuum. Only then will little matter imply a large vacuum. But this means that the existence of a vacuum is proved on the assumption that there is a vacuum (the same comment applies to Newton's observation that the path of the planets is practically undisturbed by resisting matter). Aristotle had made it clear that low density may mean fewer atoms per unit-volume and more empty space but that it may also mean low occupation by matter all over and no holes. He offers a variety of arguments for the latter alternative. I shall discuss two of them, one quantitative, the other qualitative. The qualitative argument is based on a contact-action view of motion: if a body is to move on a well-defined trajectory, it must be guided by the structure of its surroundings (this is also the point of view of general relativity). A vacuum, being a mere nothing (Democritus had identified the vacuum with non-Being) gives no guidance, hence an object cannot have a well-defined trajectory in a vacuum. The quantitative argument uses a law of motion which may be formulated as $v = F/R$, $v$ being the velocity of an object, $F$ the impressed force, $R$ the resistance. The formula gives a good approximation for motions that have stabilized under friction. Now compare two bodies with velocities $v'$ and $v''$, one of them moving in a medium, the other in a vacuum. Then, assuming any force, the above formula gives us at once the resistance of the vacuum and, as resistance is proportional to density, its density and thereby refutes the idea that there is a vacuum. Again, a vacuum seems to exclude motion though it was introduced to make motion possible. Of course, Aristotle's arguments are rather primitive when compared with modern sophistication. The point is that the arguments of his opponents were even more primitive, being either circular or irrelevant but at any rate naïvely empiristic, that Aristotle anticipated the implications of contact-action and that a vacuum, indeed, does not exist. However we approach the matter we find that we can learn a lot from Aristotle about knowledge, research, and the social implications of both. Today, when more than 30 percent of all scientists work on war-related projects, when it is taken for granted that research on recondite matters should be financed by the public, and when human existence and human nature are degraded to make them fit the most recent scientific fashions, his view that the interpretation and the use of science are a political matter is more topical than ever.

# {8} ART AS A PRODUCT OF

# NATURE AS A WORK OF ART

*1. Outline*

I WILL DISCUSS TWO CLAIMS. One is that works of art are a product of nature, no less than rocks and flowers. The other is that nature itself is an artifact, constructed by scientists and artisans, throughout centuries, from a partly yielding, partly resisting material of unknown properties. Since both claims are supported by convincing evidence, the world appears much more slippery than commonly assumed by rationalists. Intellectual generalizations around "art," "nature," or "science" are simplifying devices that can help us order the abundance that surrounds us. They should be understood as such—opportunistic tools, not final statements on the objective reality of the world.

It seems that the sciences and the arts are no longer as sharply separated as they were only thirty years ago. It is now quite fashionable to speak of scientific creativity and of the thought that enters into a work of art. Computer art, fractals, electronic music, film, debates about the role of metaphor and imagery, the whole enterprise of deconstruction have further lessened the urge for precise classifications. Yet the remaining differences are enormous.

Scientists may rhapsodize about the unity of all human efforts; they may redden with excitement when speaking about the artistic aspects of scientific research; but their tolerance vanishes when the aspects become real, enter their laboratories, and wish to be heard. And where is the scientist who would permit good, solid science money (such as a

Published in *World Futures* 40 (1994): 87–100. Section 6 of this essay was previously published as "Nature as a Work of Art" in *Common Knowledge* 1, no. 3 (winter 1992). Reprinted with kind permission. Another related essay is "Theoreticians, Artists, and Artisans" in *Leonardo* 29, no. 1 (1996): 23–28. One page in section 4 of this essay duplicates material in part II, essay 2.

small percentage of the millions that keep flowing into the Human Genome Project or of the billions promised to the Texas Supercollider) to be spent on an examination of, say, La Monte Young's music? Conversely, where is the artist, or the art commission, ready to fund a new and revolutionary science project? Even social scientists, who, after all, are dealing with people and who occasionally support the efforts of special cultures, insist on objectivity and write in a severely impersonal style.

Administrators eagerly comply. They put scientists and artists into different buildings and carefully separate their resources. We have a National Science Foundation and a National Endowment for the Humanities. The standards of both show not a trace of alleged unity of the arts and the sciences. Moreover, a large amount of philosophical rhetoric is devoted to showing that philosophical (scientific) arguments are *not* a special kind of fiction. We are not too far from the truth when asserting that the beautiful arias that are being sung about the unity of the arts and the sciences are nothing but hot air designed to conceal and to protect the strong antagonisms that still exist.

Considering these circumstances, a "lumper"—i.e., a writer who wants to unite what "splitters" want to separate—can do two things. (S)he can attack the arguments of the splitters one by one and thus weaken the intellectual resistance to unification. The procedure does not look very promising. Popular beliefs and administrative arrangements are like the hydra of legend: cut off one ugly head—and two, three, four spring up in its place. Alternatively a lumper can introduce a hydra of his/her own. This is the procedure I shall adopt.

More especially I shall argue *(first thesis)* that, like rocks and flowers, works of art are products of nature. Having done this, I shall argue *(second thesis)* that our entire universe from the mythical Big Bang via the emergence of hydrogen and helium, galaxies, fixed stars, planetary systems, viruses, bacteria, fleas, dogs to the Glorious Arrival of Western Man is an *artifact* constructed by generations of *scientist-artisans* from a partly yielding, partly resisting material of unknown properties. Both arguments are rather plausible, which shows *(third thesis)* that intellectual arguments of a general kind are uncertain allies. What we need to solve problems is experience and special pleading. So far my plan. Now on to the details!

*2. Goethe's Naturalism*    THE VIEW THAT ARTWORKS are products of nature was proposed by Goethe and elaborated by Anton von Webern in his lectures on modern music.

Commenting on Greek works of art in Italy, Goethe writes (my paraphrase):

> The magnificent works of art are at the same time magnificent works of nature produced by humans in accordance with *true* and *natural* laws.
> (J. W. Goethe, *Naturwissenschaftliche Schriften*, ed. R. Steiner [Dornach: Rudolf Steiner Verlag, 1982], 5:347)

Goethe often returns to this topic, most frequently in his *Theory of Colors* and his *Proverbs in Prose*. For example:

> Color is lawful nature working in the organ of the eye. (*Naturwissenschaftliche Schriften*, 3:88)
> Human beings insofar as they make use of their healthy senses are the largest and most precise physical instruments that can exist, and it is a great misfortune that modern physics as it were separated the experiment from the experimenter and now wants to . . . demonstrate what can be known about nature and even what she can achieve on the basis of artificial instruments alone. (5:351)
> What is beautiful is a manifestation of hidden laws of nature which without the appearance of beauty would forever remain unknown. (5:494)
> It is impossible to give an account of what is beautiful in nature and in the arts; to do this we would
> 1. have to know the laws according to which nature wants to act and acts, if she can; and
> 2. also have to know the laws according to which general nature wants to act, and acts, if she can, in the special form of human nature. (5:495)

Von Webern repeats and summarizes:

> Goethe does not recognize any essential difference between products of nature and artistic products; both are the same. What we . . . call a work of art is basically nothing but a product of general nature . . . Humans are only the vessels which receive what "general nature" wants to express. (Anton von Webern, *Wege zur neuen Musik* [Vienna: Universal Edition, 1960], 10 f.)

For von Webern, accordingly, the history of (Western) music is the history of the gradual conquest of naturally given material—the sequence of overtones. "Our major scale," Schoenberg had written in his *Harmonielehre* (Vienna: Universal Edition, 1922), 19,

> the sequence c, d, e, f, g, a, b, c, whose elements were at the basis of Greek music and of the church modes, can be explained as having been found by imitating nature; intuition and combination then helped to reconstruct the most important properties of a tone, namely, the sequence of its overtones, which we imagine as being situated simultaneously in the vertical in such a way that it now fills the horizontal, no longer simultaneously, but one overtone sounding after another.

Now if it is indeed true that works of art are products of nature in the same way in which galaxies, stars, planets, and living organisms are products of nature, then why do they look so different? The reason is that, given special conditions, the laws of nature produce special results. Physics provides many examples of this situation. Combined with laws of inertia, Newton's law of gravitation can produce falling objects, ellipses, oscillations, or chaos. In the past the different behavior of stars and stones led to the assumption that nature was divided into two large domains, one reaching from the surface of the earth to the moon, the other from the moon to the fixed stars. Similarly the difference between natural growth and artistic creation supported a separation between the sciences and the arts. The leaders of modern science removed the first dichotomy. They showed how a single set of laws, working under different conditions (on the surface of the earth; in interplanetary space), can produce qualitatively different results. Goethe and von Webern suggest that the second dichotomy be treated in the same way. I shall now discuss some consequences of this suggestion.

*3. Creativity*     THE FIRST CONSEQUENCE is that *individual creativity is considerably reduced.* If artworks are natural products then, like nature, they will change, new forms will appear, but without major contributions from isolated and creative individuals. I know that such an idea is not very popular today, when even sneezing counts as a creative act. But let us look a little more closely at the matter! Take for example the apparently very creative transition from the Homeric Gods via the God-monster of Xenophanes to the philosophical idea of Being. For Hegel this is the beginning of abstract thought. For Nietzsche the transition is the work of "giants" who communicate across an abyss populated by spiritual midgets *(Gezwerge)*. More prosaic writers like Mircea Eliade or W. K. C. Guthrie speak of a fundamental discovery, made by individuals of superior intelligence. What did really happen?

Gilbert Murray, the great classical scholar, gives us a hint. The Greek Gods started as local powers. They lived in well-defined surroundings, on a mountain, for example, or in the fields. Journeys of discovery, the search for colonies, warlike enterprises then made the travelers acquainted with new divinities which in some respect differed from the familiar Gods but resembled them in others. Occasionally they had even the same name. The travelers emphasized the similarities and disregarded or overlooked the differences. As a result the Gods lost in individuality and humanity but gained in power, for their radius of action was now vastly increased. The changes occurred slowly and gradually. Many people were affected, but without consciously contributing to the process.

There were analogous developments in other fields. Buying and selling started as an exchange of gifts: an object that was not only useful but had personal memories attached to it was exchanged for another object of a similarly complex nature. Aesthetics, family history, and practical value were closely connected. Then intermediate objects, a "currency," entered the process. At first these objects were intrinsically valuable (iron rods, or silver coins); later on they received value from the mode of circulation. Again a property of things—their "value"—got detached from personal elements and became more abstract.

Democratization fits right in to this pattern. In early times cities were ruled by power families. Politics was a family concern; it was guided by loyalty, friendship, and personal obligations. Slowly this situation gave way to general definitions of rights and duties. The change was not intended. It was the unintended side effect of special arrangements (Solon, Cleisthenes) designed to solve special political problems. The solutions coalesced—and democracy was the result. Even language gradually lost in content: "Words become impoverished in content, they turn into formulae, become empty and one-sided."[1] The new social critics, the philosophers did not oppose these trends. They praised them, acted as if they had started the affair and raised the result, conceptual poverty, to a principle. They were parasites of changes that had occurred without any creative interference on their part.[2]

A second example makes the situation even clearer. Simon Stevin, a Dutch scientist of the late sixteenth and early seventeenth centuries, wanted to prove that a chain put around a wedge will be in equilibrium if and only if the weights of the sections lying over the sides of the wedge are related to each other as are the lengths of these sides. Assuming that the chain is closed and that its weight is equally distributed over all its sections, he argued as follows: if the chain moves, then it must move forever, for every position is equivalent to every other position; if, on the other hand, it is without motion, then it will also remain without motion, i.e., it will be in equilibrium. The first possibility can be excluded—there are no perpetual motions. In the second case we can remove the lower part of the chain, because of its symmetry—and the result becomes obvious.

How did Stevin know that the chain would remain at rest and that perpetual motion was impossible? Was he creative? Did he creatively suggest a bold hypothesis? Ernst Mach, who analyzed the case, denies this. Stevin, he says, had adapted to his surroundings and moved in his imagination as the surroundings moved in reality. It would have been most surprising to see a chain that suddenly starts moving. Why? Because a plethora of data had turned into an instinct, which from then on guided the thinker. It is the nature of this instinct or, in other words, *it is nature as it manifests itself in a particular person that shows the way, not a mysterious "creativity."* Mach applied the lesson to our knowledge of numbers. "It is often the case," he wrote in *Erkenntnis und Irrtum* ([Leipzig: Barth, 1917], 327)

> that numbers are called "free creations of the human mind." The admiration for the human mind which is expressed by these words is quite natural

---

1. Kurt von Fritz, *Philosophie und sprachlicher Ausdruck bei Demokrit, Platon und Aristoteles* (reprint, Darmstadt: Wissenschaftliche Buchgesellschaft, 1966), 11.

2. Philosophers opposed democracy not because it was general and abstract, but because it was not general enough. After all, people were still allowed to deflect the democratic process by their own idiosyncratic demands.

when we look at the finished, imposing edifice of arithmetic. Our understanding of these creations is, however, furthered much more when we try to trace *their instinctive beginnings* and consider the circumstances which produced the need for such creations. Perhaps we shall then realize that the first structures that belong to this domain were unconscious biological structures which were *wrested from us* by material circumstances and that their value could be recognized only after they had appeared.

*4. The Scattered Unity* THE TWO EXAMPLES MAKE IT CLEAR that in-
*of Human Efforts* ventions in metaphysics and in the sciences are
not isolated acts of solitary thinkers but are
linked to nature in many ways. There is novelty—but it is a common feature of natural processes. And psychological studies of problem solving show indeed that decisive elements of a difficulty frequently "arrange themselves" in a way that is independent of personal wishes and efforts.

A second consequence of Goethe's approach is *that human activities though closely related to each other—they obey basic natural laws—are scattered and diverse: the diversity of human idiosyncrasies modifies the laws in many ways.* A brief look at the sciences and the arts confirms this second consequence.

Music for a long time was regarded as a way to knowledge. For Saint Augustine the perfect chords represented truth in a way inaccessible to human reason. For Grosseteste music, not physics, revealed the innermost structure of things. Later, when the sciences had separated from the arts and the arts in turn had separated from the crafts, some writers regarded music as a paradigm of artistic excellence. "All art aspires to the condition of music," wrote Walter Pater.[3] Kant, on the other hand, regarded music, which, for him, was "more pleasure than culture," as the lowest art form and separated it from all epistemic claims.[4]

Painting for Plato was deceptive and unreal. Painting, sculpture, even architecture were excluded from early university curricula. Painters belonged to the guild of sign painters, wall painters, or apothecaries (who prepared their colors). After the discovery of perspective and Leon Battista Alberti's vigorous propaganda for it, painters divided into those "who knew" and others who preferred to follow tradition. Aided by ar-

3. Pater, "The School of Giorgone," in *The Renaissance* (London: Macmillan and Co., 1894).
4. *Kritik der Urteilskraft*, sec. 53.

chitects and sculptors, the "theoreticians" soon founded academies.[5] Then painters rejected what had given them substance, art critics started emphasizing the uniqueness of individual works of art, and some artists pretended to live by creativity and/or accident alone. That changed not only the philosophical evaluation of the arts, but also their content: there is hardly any connection between Raphael and Jackson Pollock.[6] General distinctions between the arts and the sciences existed since antiquity, but the reasons differed and so did the distribution of subjects among the two categories.[7] Thus some seventeenth-century writers asserted that, while ancient science had been overcome by the science of Galileo and Descartes, the ancient arts, poetry especially, still reigned supreme and were therefore different in nature from scientific products.

What is true of the arts is true of the sciences. Twentieth-century philosophy of science for a long time identified science with physics and physics with relativity and elementary particle physics; space, time, and matter, after all, are the basic ingredients of everything. A uniform *conception of knowledge* separated SCIENCE from other enterprises and gave it substance. A look at *scientific practice* tells a different story.

For here we have scientists such as S. Luria who tie research to events permitting "strong inferences" and favor "predictions that will be strongly supported and sharply rejected by a clear-cut experimental step."[8] According to Luria, decisive experiments in phage research had precisely this character. Scientists of Luria's bent show a considerable "lack of enthusiasm in the 'big problems' of the Universe or of the early

5. The theoretical trend is represented by, e.g., Leon Battista Alberti's essay *On Painting*, tr. J. R. Spencer (New Haven: Yale University Press, 1966); cf. also Joan Gadol, *Leon Battista Alberti* (Chicago: University of Chicago Press, 1973). Tradition's bible was Cennino Cennini, *The Craftsman's Handbook*, tr. D. V. Thompson Jr. (New York and London: Dover Publications, 1959). For the rise and the fate of academies cf. N. Pevsner, *Academies of Art, Past and Present* (Cambridge: Cambridge University Press, 1940).

6. Interestingly enough there may be a phase difference between a style and its philosophical evaluation, and the latter may be ignorant of the style to which it applies. Cf. Carl Dahlhaus, *Klassische und romantische Musikästhetik* (Laaber: Laaber Verlag, 1988), esp. chap. 11.

7. According to Plato (*Republic*, bks. 7 and 10), music has a practical and a theoretical side (and in this respect is similar to arithmetic) while painting has neither (and is therefore both useless and without epistemic merit). Plato also points out that astronomy is still lacking in theory and in this respect inferior to music.

8. S. E. Luria, *A Slot Machine, a Broken Test Tube* (New York: Harper and Row, 1985), 115.

Earth, or in the concentration of carbon dioxide in the upper atmosphere," all subjects "loaded with weak inferences."[9] In a way they are continuing the Aristotelian approach, which demands to remain in close contact with experience and objects rather than following plausible ideas to the bitter end.[10]

However, this was precisely the procedure adopted by Einstein, by students of the stability of the planetary system between Newton and Poincaré, by the early proponents of the kinetic theory, and by almost all cosmologists. Einstein's first cosmological paper was a purely theoretical exercise containing not a single astronomical constant. The subject of cosmology itself for a long time found little respect among physicists. Hubble, the empiricist, was praised—the rest had a hard time:

> Journals accepted papers from observers, giving them only the most cursory refereeing whereas our own papers always had a stiff passage, to a point where one became quite worn out with explaining points of mathematics, physics, fact and logic to the obtuse minds who constitute the mysterious anonymous class of referees, doing their work, like owls, in the darkness of the night. (F. Hoyle, "Steady State Cosmology," in *Cosmology and Astrophysics*, ed. Y. Terzian and R. M. Bilson [Ithaca and New York: Cornell University Press, 1982], 21)

"Is it not really strange," Einstein wrote in one of his letters to Max Born, "that human beings are normally deaf to the strongest arguments while they are always inclined to overestimate measuring accuracies?"[11]—but just such an "overestimating of measuring accuracies" is the rule in spectroscopy, celestial mechanics, genetics, and even in subjects such as demography, epidemiology, some parts of anthropology—and so on. And there exist still other views about the relation between theory, experiment, and fact: *science does not contain one epistemology, it contains many.*

Moreover, none of these epistemologies is tied to specific subjects. "Aristotelians" may abound among bird watchers and phage enthusiasts; however, they also turn up in cosmology (examples: Heber Curtis in the "Great Debate" with Shapley; Ambartsumian; Halton Arp and his collaborators), hydrodynamics (Ludwig Prandtl, for example), quantum

9. Ibid., 119.

10. Luria reports that Fermi had little sympathy for speculative theories such as the general theory of relativity. (The same was true of Michelson, Rutherford, and even Planck.)

11. *The Born-Einstein Letters* (London: Macmillan, 1971), 192.

theory (cross-section enthusiasts), thermodynamics, mechanics (engineering mechanics)—you name it.

Now the interesting thing is that many of these different approaches were successful in the sense that they produced acceptable facts, laws, and theories.[12] But this means that, being constructed in different ways, different scientific knowledge claims cannot be easily combined and that *the idea of a coherent "body of scientific knowledge" is a chimera.* [13]

I conclude that terms such as SCIENCE and ART are temporary collecting bags containing a great variety of products, some excellent, others rotten, all of them characterized by a single label. But collecting bags and labels do not affect reality; they can be omitted without changing what they are supposed to organize. What remains are events, stories, happenings, results which may be classified in many ways but which are not divided by a lasting and "objective" dichotomy. This confirms the second consequence of Goethe's approach.

We can go further and assert that *both scientists and artists (artisans) learn by creating artifacts.* I shall illustrate the assertion by an example from architecture.

*5. Early Gothic*     Eugène Emmanuel Viollet-le-Duc, the nine-
*Artifacts*           teenth-century architect, archaeologist, and
                      writer, assumed that the medieval masters strove
for efficient structures and had ways of controlling the consequences of their actions. Summing up his research he suggested that architects

12. Even highly implausible approaches have led to success. An example is Maxwell's calculation of the viscosity of gases. For Maxwell this was an exercise in theoretical mechanics, an extension of his work on the rings of Saturn. Neither he nor his contemporaries believed the outcome—that viscosity remains constant over a wide range of density—and there existed contrary evidence. Yet more precise measurements confirmed the prediction and thus, indirectly, the kinetic approach. Cf. *The Scientific Papers of James Clerk Maxwell*, ed. W. O. Niven (1890; reprint, New York: Dover Publications, 1965), 377 ff. For more recent conflicts between physical common sense and mathematical theory ending in a triumph of theory cf. G. Birkhoff, *Hydrodynamics* (New York: Dover Publications, 1955), secs. 20 and 21.

13. Details in part II, essay 2, of this volume. The idea that "peripheral" knowledge claims can be reduced to "more fundamental ones" and, ultimately, to elementary particle physics, which underlies the idea of a coherent body of scientific knowledge, is a metaphysical desideratum, not a fact of scientific practice. For details cf., e.g., Nancy Cartwright, *How the Laws of Physics Lie* (Oxford: Clarendon Press, 1984).

should acquire practical experience, learn the "inexorable" and objective laws of statics, and pay only slight attention to artistic forms.[14]

This would be sound advice if (a) the practical experience needed for the realization of artistic forms could be obtained independently of studying these forms; if (b) the "inexorable laws of static" could be found in the same manner, i.e., independently of any artistic enterprise; and if (c) practical experience combined with the laws sufficed for defining the form of any building.

None of these assumptions is correct. Assumption (c) is refuted by the great variety of styles that existed at different periods and in different regions of the European continent. Adding economic constraints to explain, say, the transition from solid walls with small windows to the evanescent materiality of Gothic cathedrals overlooks the fact that there were other ways of saving time, material, and work such as smaller churches, open-air meeting places, no churches at all. None of these alternatives was adopted, which shows that noneconomic requirements were at work. And indeed—most art historians now point out that the Gothic innovations arose from a special view of Divine Nature and of possible approaches to it.[15]

According to Pseudo-Dionysius Areopagita, an otherwise unknown Neoplatonist writing about A.D. 500, ultimate reality (God, Being) is ineffable. Trying to grasp it directly we face darkness, silence, nothingness. But Being does not remain self-contained. It expands and, while expanding, creates hierarchies, first of light, then of lower forms down to coarse matter. Matter is far removed from the primary cause; still, it contains traces of it. Abbot Suger of Saint Denis, one of the most energetic proponents of the new style, believed that the traces could be amplified by precious stones, gold, and shining objects and, so he hoped, by special arrangements of light and matter. This hope was one of the driving forces behind the development of the Gothic style.

To realize his hope, Suger had to overcome the resistance of matter. Matter was (and still is) only imperfectly known. Experience rests on earlier work and is changed by responding to new problems. Theories are

14. Cf. Robert Mark, *Experiments in Gothic Structure* (Cambridge: MIT Press, 1982), 11.

15. Survey and references in Otto van Simson, *The Gothic Cathedral* (New York: Bollingen Foundation, 1962).

either speculative or empirical. In the first case they can be modified by further thought. In the second case they are as restricted as the experience that supports them. The theories available at Suger's time were fragmentary, inferior by far to the experience of architects and masons and unrelated to Suger's intentions.[16] New experiences were needed to get ahead.

New experiences are needed even today if we want to judge the results of Suger's efforts and the efficiency of the Gothic style in general. One reason is that the theories of elasticity (for example) which arose in the eighteenth and nineteenth centuries quite intentionally kept away from practical matters.[17] Another reason is that modern attempts at a theoretical evaluation are "at least half a century too late. Modern physics has veered away from the study of analytic structural mechanics, leaving its development in the hands of research engineers."[18] Thus the resistance of matter to (metaphysical, theological, technological, artistic) transformation is in fact being determined by a procedure that takes each case separately and judges it on its own merits. We can say that matter responds positively to some approaches and that it frustrates others; we cannot say that, taken together, the approaches reveal a stable and "inexorable" nature of the elements used. That—modern scientists say—is done by theory. Nature, they assert, is not a patchwork of practical results, it is what is being described by overarching theoretical principles. Is a nature so defined immune to human interference? The second thesis of section 1 denies that it is.

*6. Nature as an Artifact*    ACCORDING TO THIS THESIS,[19] *nature as described by our scientists is an artifact that is constantly being enlarged and rebuilt by them.* In other words: our entire universe from the mythical Big Bang via the emergence of hydrogen, helium, galaxies, fixed stars, planetary systems, viruses,

---

16. Cf. the material and literature in E. Grant, ed., *A Source Book in Mediaeval Science* (Cambridge: Harvard University Press, 1974).

17. Cf. the historical introduction in A. E. H. Love, *Treatise on the Mathematical Theory of Elasticity* (Cambridge: Cambridge University Press, 1927).

18. Mark, *Experiments in Gothic Structure*, 13.

19. In this section I have made use of formulations first published in *Common Knowledge* 1, no. 3 (winter 1992).

bacteria, fleas, dogs to the Glorious Arrival of Western Man has been constructed by generations of *scientist-artisans* from a partly yielding, partly resisting material of unknown properties.

The thesis seems to be trivially false. The universe is much bigger than humans and it existed long before they appeared in it; hence it could not possibly have been built by some of them.

But it is not unusual for artisans to misjudge the implications of their activity. Platonists assume that numbers are independent of the human race. They existed before the first human appeared and they will endure after the last human has left the earth. Yet many people now believe that numbers emerged from complex social activities and seem timeless because they have become part of the hardware of language. Looking upward on a clear evening, do we not perceive a blue sphere that seems to cover all we can see? And has this perceptual fact (which was much more obvious in the ancient Near East than it is now in our cities) not been taken as proof of the eternity of the heavens? And yet we are being told that there is no sphere whatsoever, only a gaping void filled here and there with small amounts of matter and radiation. Divinities play a large role in many cultures; they exceed humans in power, existed before them, may have created them, and are often perceived by them. But a large percentage of the Educated Few now regards Gods as projections, i.e., again as (unconscious) artifacts. Thus it is quite possible for artifacts to have features that, when taken as real, seem far more powerful than their creators.

How did this mistake arise? How did it happen that mundane matters were blown up to such an extent that they seemed to exceed their creators in size, power, and duration?

Platonism started as a philosophy or, if you like, as a vision, or a myth. Those who believed in it were not easily deflected by objections. One might say that they were rather dogmatic. However, they were dogmatic in an interesting way. They did not just sit on their myth; they did not abandon it either, despite the criticisms they received; *they put it to work*. Like artisans, they made their vision produce tangible results. A large amount of modern mathematics, Cantor's unusual speculations included, was created by their efforts. After that, the task of the opponents became much more difficult. They no longer faced a short and, perhaps, slightly ridiculous *story* ("I can see horses, dear Plato," said Antisthenes, "but not THE HORSE") but a complex assembly of challenging and useful *objects*. The modern debate between Platonists and those who regard mathematics as a human invention is as much about these objects (integers, lines, points, irrational numbers, transcendent numbers, classes, transfinites, and so on) as it is about metaphysics: for Platonists these objects were found, while constructivists assert that they were constructed, all of them, despite the obvious limitations of human thought and actions.

Empirical research has exactly the same features. As in mathematics, there existed a variety of views, or visions, about the nature of knowledge and the structure of the world. God often played an important role. For Saint Thomas (and Descartes) He was a rationalist of sorts. He guaranteed the eternal and inexorable truth of universals and scientific principles (*Summa Theologica*, question 8, art. 4). Duns Scotus and William Ockham emphasized the will of God. The will of God, they said, is unfathomable. All we can do is record the results of divine action, arrange them in a convenient way, and hope for the best: an extreme empiricism was justified by theological arguments.[20] Newton rejected the God of Descartes and Spinoza. For him God was a person showing concern and demanding respect, not an abstract principle. Newton also assumed— and he had empirical reasons for doing so—that God from time to time checked the planets and reset their motions; He was a much-needed ordering force in the universe.[21] Kepler thought that reacting to special conditions the telluric soul caused earthquakes, floods, and atmospheric aberrations. Tycho Brahe, his great predecessor, still believed in miracles. Such assumptions seem strange today after science has trimmed most facts and declared others to be subjective and, therefore, irrelevant. However, they were perfectly adapted to the empirical knowledge of the time.

Comets appeared, grew to monstrous size, and faded away; there were meteors, haloes, triple suns, new stars, and other ominous events. Strange geological shapes, earthquakes, volcanic eruptions proved that nature could not be fit into simple patterns, while malformations in

20. G. de Ockham, *Opera philosophica et theologica*, vol. 1, *Scriptum in librum primum sententiarum ordinatio: Prologus et distinctio prima*, ed. G. Gál and S. F. Brown (Saint Bonaventure, N.Y.: Franciscan Institute, Saint Bonaventure University, 1967), 241.

21. Frank Manuel, *The Religion of Isaac Newton* (Oxford: Clarendon Press, 1974). Cf. also query 31 of Newton's *Opticks*.

plants, animals, and humans made it difficult even to think of general biological laws. Naturalists praised the incredible variety of life as a sign of the richness of God's creative powers, while some early psychologists (not yet classified by that name) found and described a veritable snakepit of behavioral aberrations (outstanding example: the descriptive sections of the *Malleus Maleficarum*). The early Chinese seem to have taken this situation at face value. They recorded facts, emphasized unusual appearances, concentrated on description, and eschewed far-reaching generalizations. They were true empiricists. So was Aristotle. He divided the world into sections, each with its own principles, admitted deviations from the norm ("natural is what applies universally *or in most cases"—De partibus animalium* 663b27 ff., my emphasis) and used general notions only to serve the whole. And I have already mentioned how Tycho Brahe, Kepler, and Newton dealt with the matter. Yet some leading Western theoreticians, Descartes, Galileo, and Leibnitz among them, disregarded phenomena and postulated "universal and inexorable laws." In a way they repeated what Platonists had done with numbers. But while the Platonists faced only a philosophical opposition, these writers had to contend with experience as well. Their myth was not only implausible, it was also empirically absurd. Did they withdraw? No. They stuck to their myth, introduced new facts, and crushed the opposition with their weight.

Simplifying matters, we may say that they changed existing knowledge in two ways. They emphasized experiment over observation and they considerably extended the use of mathematical formalisms. In both cases *they replaced natural processes by artifacts.*

Take a simple scientific measurement such as weighing with a precision scale. Note that already the older scales were carefully constructed. They were enclosed in containers to exclude drafts; temperature was kept constant; possible effects of magnetism and electricity were eliminated; there were corrections for buoyancy, for impurities of the substance weighed, and for other "disturbing" effects. The result of the operations was a property, the "mass" of the product, which is a measure of the force that needs to be applied to it to overcome, i.e., to keep it from falling in, a given gravitation field. Modern elementary particle experiments have pushed this aspect to an extreme. Here we have entire cities, watched around the clock like submarines or sensitive sections of the Pentagon, their intestines protected from undesirable influences while their active parts produce events that cannot be seen, not even in principle, but are recorded and interpreted by complex and highly sophisticated instruments.

Next, consider language. When Newton applied his laws of gravitation to the planetary system, he used Euclidean geometry. That was already a step away from the practical geometry of masons and carpenters. Still, every assumption he made could be visualized and controlled by the imagination. Trying to remove discrepancies, his successors introduced algebraic methods. The discrepancies remained until Laplace, more than a hundred years after the *Principia,* found a solution. Then it was discovered that Laplace's series diverged after having converged and that there was no other way of getting quantitative results. Poincaré, undeterred, developed new (topological) methods, which have remained in place until today. For over two hundred years Newton's law of gravitation could not deal with the stability of the planetary system. Yet nobody divided the world into parts and declared the law to be successful only in some of them. The belief in the universal validity of Newtonianism was retained until new types of measurement and a new mathematics enabled scientists to solve this most difficult problem.

Examples such as these show very clearly that modern science uses artifacts, not Nature as She is. Can we infer that the final product, i.e., nature as described by our scientists, is also an artifact, that nonscientific artisans might give us a different nature and that we therefore *have a choice* and are not imprisoned, as the Gnostics thought they were, in a world we have not made?

It seems that we cannot. Granted—experiments interfere with nature and their results are recorded and processed in rather "unnatural" ways. But the interference has its limits. Nature is not something formless that can be turned into any shape; it resists and by its resistance reveals its properties and laws. Besides, experiments do not just interfere, they interfere in a special way. They eliminate disturbances, create strong effects, and enable us to watch the underlying machinery of nature undistorted and enlarged. Having concluded our investigation, we can therefore forget about the experiments and speak about nature as it is independently of all disturbances. Not only that—reapplying our instruments and using the knowledge we have gained, we can produce new effects and reshape our surroundings. Modern technology and modern medicine show to what extent we have mastered the laws that govern the universe.

This popular argument, which seems to be an inseparable companion of scientific research, rests on the assumption that scientists proceed in a uniform way and that their results form a single coherent picture; whoever does research and whoever performs experiments runs into the same type of facts and the same set of laws. In section 4, I gave reasons for rejecting this assumption: science is not one thing, it is many; and its plurality is not coherent, it is full of conflict. Even special subjects are divided into schools. I added that most of the conflicting approaches with their widely different methods, myths, models, expectations, dogmas *have results.* They find facts that conform to their categories (and are therefore incommensurable with the facts that emerge from different approaches) and laws that bring order to assemblies of facts of this kind. But this means that *being approached in different ways Nature gives different responses* and that projecting one response onto it as describing its true shape is wishful thinking, not science.

Let us discuss the matter in more general terms. The success of a particular research program, say, molecular biology, or of a particular project, such as the Human Genome Project, can be explained in at least two ways. First way: the procedures (experiments, ideas, models, etc.) that are part of the program and that strongly interfere with Nature *reveal how Nature is* independently of the interference. Second way: they *reveal how Nature responds to the interference.* Adopting the second way, we say that the world as described by scientists is the result of a complex exchange between Nature as She Is In and For Herself—and this lady we shall never know—and inquisitive research teams including, possibly, the whole subculture that supports them. Which way shall we adopt? And why would we prefer it?

I already mentioned one reason in favor of the second way: the plurality inherent in science itself. Scientists adopting different myths and using corresponding procedures get respectable results. They succeed, which means that their courtship of Nature did not remain unanswered.

A second reason is provided by more recent discoveries in anthropology, in the history of astronomy, medicine, mathematics, technology and especially by the findings of theologians, ecologists, medical people engaged in developmental aid. These discoveries and findings show that nonscientific cultures provided acceptable lives for their members and that the imposition of Western ideas and practices often disturbed the delicate balance with nature they had achieved. Nonscientific cultures are (were) not perfect—no way of living is—but they were often better than what succeeded them. Finally, there is the quantum theory. It is one of the best-confirmed theories we possess and it implies, in a widely accepted interpretation, that properties once regarded as objective depend on the way in which the world is being approached. Taking all this into consideration, I conclude that the second thesis makes lots of sense: *nature as described by our scientists is indeed an artifact built in collaboration with a Being sufficiently complex to mock and, perhaps, punish materialists by responding to them in a crudely materialistic way.*

Now at this point it is important not to fall into the trap of relativism. According to the second way, mentioned above in the present section, nature as described by scientists is not Nature In and For Herself, it is the result of an interaction, or an exchange, between two rather unequal partners, tiny men and women on the one side and Majestic Being on the other. Not all exchanges produce beneficial results. Like unfit mutations, the actors of some exchanges (the members of some cultures) linger for a while and then disappear (different cultures have much in common with different mutations living in different ecological niches). The point is that there is not only one successful culture, there are many, and that their success is a matter of empirical record, not of philosophical definitions: an enormous amount of concrete findings accompanies the slow and painful transition from intrusion to collaboration in the field of development. Relativism, on the other hand, believes that it can deal with cultures on the basis of philosophical fiat: define a suitable context (form of life) with criteria etc. of its own and anything that happens in this context can be made to confirm it. As opposed to this, real cultures change when attempting to solve major problems and not all of them survive attempts at stabilization. The "principles" of real cultures are therefore ambiguous and there is a good sense in saying that *every culture can in principle be any culture.* Applied to science this means that the artwork of science, in many respects, resembles Kurt Schwitters's fantastic productions—there are recognizable details, there are features that seem devoid of sense, and there is the general invitation to add to the thing and in this way to change the appearance of the whole. Scientific nature, too, is partly comprehensible, partly nonsensical; it can be extended, changed, supplemented with new ideas, habits, pieces of culture thus bringing to light other and perhaps more gentle aspects of Nature and, with that, of ourselves. Here progressive artists can play an important role. Rationalists—and that includes many scientists and philosophers—like to nail things down. They are confused by change and they cannot tolerate ambiguity. But poets, painters, musicians cherish ambiguous words, puzzling designs, nonsensical movements, all instruments which are needed to dissolve the apparently so rigid and objective nature of scientists, to replace it by useful and changing *appearances* or *artifacts* and in this way to give us a feeling for the enormous and largely unfathomable powers that surround us.

| | |
|---|---|
| *7. Lessons to Be Learned* | AND WITH THAT I am back at the topic from which I started—the relation between artworks and the world. As I presented it, this topic leads |

right into one of the most pressing problems of today—the side effects of a ruthlessly "objectivistic" approach. Objectivism certainly is not the only problem. There are the rising nationalisms, the greed, stupidity, and uncaring attitude of many so-called world leaders, in politics, religion, philosophy, the sciences, all this accompanied by a general thoughtlessness that seems satisfied and even pleased with the repetition of tepid generalities. Arguing for two theses which seem to be in conflict, I tried to undermine this thoughtlessness and to show how easy it is to find evidence now for the one, now for the other point of view: *the world is much more slippery than is assumed by our rationalists* (this is the content of the third thesis). But there is also a positive result, namely, an insight into the abundance that surrounds us and that is often concealed by the imposition of simpleminded ideologies. Many aspects of this abundance have been studied, by scientists, development workers, liberation theologians, they have been given shape by painters, poets, musicians, and even the most downtrodden inhabitants of our globe have made their contribution, provided one asked them, not in general terms, but with reference to things right before their nose, and respected their answers. It is true that allowing abundance to take over would be the end of life and existence as we know it—abundance and chaos are different aspects of one and the same world. We need simplifications (e.g., we need bodies with restricted motions and brains with restricted modes of perception). But there are many such simplifications, not just one, and they can be changed to remove the elitism which so far has dominated Western civilization.

# {9} ETHICS AS A MEASURE OF SCIENTIFIC TRUTH

## COMMENTS ON FANG LIZHI'S PHILOSOPHY OF SCIENCE

I N A SPEECH that was read in his absence in Washington in November 1989 (and was translated into English for publication in the *New York Review of Books* of December 21, 1989, "Keeping the Faith") Fang Lizhi, the Chinese astrophysicist and dissident, argued for

> universally applicable standards of human rights that hold no regard for race, language, religion, or other beliefs.

The advance of civilization, he said,

> has largely followed from the discovery and development of just such universally applicable concepts and laws. Those who reject the idea that science has universal application were, in fact, doing nothing more than demonstrating their fear of modern culture.

A country that wants to catch up with the modern world must therefore

> change by absorbing those aspects of modern civilization, especially science and democracy, that have proven both progressive and universal.

Such aspects are not only necessary for a rewarding life, they correspond to basic features of the universe. Here, in the universe,

> the first principle is called the "cosmological principle." It says that the universe has no center, that it has the same properties throughout. Every place in the universe has, in this sense, equal rights. How can the human race

Published in W. R. Shea and A. Spadafora, eds., *From the Twilight of Probability* (Canton, Mass.: Science History, 1992), 106–14, © 1992 Science History Publications, USA. Reprinted with kind permission. First published in Italian as "L'etica come misura di verità scientifica," *Iride* 7 (1991): 68–76; and in Spanish as "Etica como medida de verdad científica," in *Feyerabend y Algunas Metodologías de la Investigación*, ed. A. M. Tomeo (Montevideo: Nordan, 1991). One page toward the end of this essay duplicates material in part I, essay 3, section 2, in part II, essay 4, section 4, and in part II, essay 7.

that has evolved in a universe of such fundamental equality, fail to strive for a society without violence and terror? And how can we not seek to build a world in which the rights due every human being from birth are respected?

Fang concluded his talk with an invocation: "May the blessings of the universe be upon us!"

IN WHAT FOLLOWS I shall criticize Fang. I shall try to show that his assumptions give a false impression of the sciences and are liable to endanger the very kind of life he tries to defend. I don't want to interfere with his politics. I do not know the situation and abstract celebrations of freedom are not my cup of tea. After all, who knows if being in China, knowing the language, the concerned parties, and the bystanders, peasants included, I might not suggest entirely different actions and policies. Nor do I want to imply that wrong and even inhumane ideas might not be the right way of changing an inhumane system. Ideas of humanity change. Is it inhumane to save the life of an enemy? Yes, if it means that he will soon be able to do what he does best—rape women and kill children. Right now all these matters do not concern me. What concerns me is a point of view that is shared by Fang, by some of his followers, and by many Western admirers of the monster "science." This point of view contains a totalitarian element. It is good to know this, even if one should decide, for tactical reasons, to retain it for a few more years.

THE IDEA THAT THE UNIVERSE sets an example for humans but that humans go their own way and suffer as a consequence is not new. It occurs in the *Hopi Genesis*, in Plato's *Timaeus*, and in many other stories. Fang accepts the general outline but changes the specifics. His universe is "democratic"; it has no preferred places; and it supports universality over what is specific and irregular. Human suffering can be alleviated, violence and terror removed by acting accordingly. Or, as Fang expresses it, "by absorbing those aspects of modern civilization that have proved both progressive and universal."

Fang recognizes that desirable changes hardly ever occur by themselves. It needs wisdom and much research to discover the true nature of the universe, determination to follow it, and power, the power of dissidents included, to remove obstacles. But wisdom and research have failed, power has been misused, and terror, not peace and happiness, has arisen from attempts to imitate an alleged cosmic order. Let us therefore examine Fang's advice a little more carefully. More especially, let us examine the idea of universality it contains. Science is universal, says Fang; what does that mean and what follows if it is?

It cannot mean that scientists in Chicago and Beijing *accept* the same basic ideas and *use* the same basic methods. Fang speaks about the properties of concepts and laws (they are "universally applicable"), not about the actions of scientists. Besides, scientists disagree about fundamental matters, even within the confines of a single country. For example, there are scientists for whom Fang's "cosmological principle" is a simplifying artifice, not a cosmological fact.

Nor can the universality of science mean that all scientific laws are universally true and all methods universally applicable. Many laws, methods, disciplines are restricted to special domains. For example, the specific laws of hydrodynamics are neither valid nor thought to be valid in elementary particle physics.

However, science does seem to be universal in the more restricted sense that it contains universal principles, i.e., principles that apply to everything. Examples would be the second law of thermodynamics, energy and momentum conservation, the limiting nature of the velocity of light, and the basic laws of quantum mechanics. Is this impression correct? And what follows if science is indeed universal in the sense just described?

For Fang, the universality of science is closely connected with its uniqueness, i.e., with the claim that there is only one science and only one type of genuine knowledge. But the fact that a discipline contains unrestricted principles does not exclude other disciplines with other and equally unrestricted principles in their repertory. Arithmetic does not exclude geometry, and phenomenological thermodynamics for a long time developed side by side with mechanical principles. True, alternatives may merge or be reduced by elimination: science *occasionally* (though hardly ever) speaks with a single voice. *Permanent* uniqueness, however, is not a fact. It is an *ideal,* or a *metaphysical hypothesis.*

Now I have no objection to metaphysical hypotheses. On the contrary, I would say that science is impossible without them. My reason is that no interesting scientific idea has ever had a clean bill of health. Theories are beset by empirical and logical difficulties when they are first proposed, and they continue to be in trouble long after they have become part of scientific common sense. Rejecting them because of their imperfections would be the end of science. Retaining them despite their faults and in the face of more successful (but less interesting) alternatives means making conjectures that transcend experience and are in this sense metaphysical. Some examples will explain what I mean.

Darwin's theory conflicted with the fact that life seemed to start in post-Cambrian times. Leading professionals, Murchison among them, inferred some form of creation. Darwin persisted: life did start earlier, but its traces had not yet been found.

Einstein's theory of special relativity clashed with evidence produced only one year after its publication. Lorentz, Poincaré, and Ehrenfest withdrew to a more classical position. Einstein persisted: his theory, he said, had a wonderful symmetry and should be retained. He gently mocked the widespread urge for a "verification by little effects."

Schrödinger's first wave equation was adapted to the most recent view on space and time (it was "relativistically invariant") but led to incorrect predictions. Schrödinger stuck to his basic assumption (wavelike character of elementary processes), combined it with older space-time views, and got correct results. This is a very interesting situation: the "better" theory fails, the "inferior" theory succeeds.

Newton's mechanics could not account for the stability of the planetary system. Newton himself thought that God periodically put the heavenly machine in order. It took about 150 years until a reasonable solution was found—only to be proved impossible a few decades later. Still, scientists did not despair. They chose a new approach which so far seems to do the trick.

No; metaphysics is not the problem. The problem in our present case is that a metaphysical principle is presented as a well-established fact and that people are invited to follow a science so distorted. For example, they are invited to "absorb" an allegedly unique Western science and to abolish other forms of knowledge.

Metaphysics also affects the matter of universality. We can assume that for Fang the universality of a principle means that it corresponds to universal features of an observer- and history-independent world. But such a correspondence is not obvious. What the evidence tells us is that having approached the world or, to use a more general term, Being, with concepts, instruments, interpretations which were the often highly accidental outcome of complex, idiosyncratic, and rather opaque historical developments, Western scientists and their philosophical, political, and financial supporters got a finely structured response containing quarks, leptons, space-time frames, and so on. The evidence leaves it open if the response is the way in which Being reacted to the approach, so that it reflects both Being and the approach, or if it belongs to Being independently of any approach. Realism assumes the latter; it assumes that a particular phenomenon—the modern scientific universe and the evidence for it—can be cut from the development that led up to it and can be presented as the true and history-independent nature of Being. The assumption is very implausible, to say the least.

For are we really to believe that people who were not guided by a scientific worldview but who still managed to survive and to live moderately happy and fulfilling lives were the victims of an illusion? They had definite ideas about the world and its constituents. They noticed, reacted to, and arranged their lives around all sorts of entities, Gods, saints, demons, spiritual elements of matter among them. The space they inhabited never obeyed a "cosmological principle"—it had collective centers and, assembled around them, numerous individual centers. Yet they were by no means more disoriented than we are; on the contrary, their lives were occasionally less scattered, aimless, and cruel than our own. Is it plausible to assume that all this was a grandiose mistake? Or is Being perhaps more yielding than our materialists are willing to concede? Were the Gods, saints, demons, souls (of people and animals), and the centrally structured spaces that played such an important role in their lives perhaps the way in which Being received their approach so that, given this approach, they were as real as elementary particles are supposed to be today? The evidence assembled by scientists does not conflict with the latter interpretation. On the contrary, one of the most fundamental twentieth-century theories, the quantum theory, suggests that the familiar spatiotemporal properties of elementary particles are not inherent in them but emerge as the result of special interactions. Yet, many scientific realists remain unconvinced. Those of them who pay attention to the results of anthropologists and classical scholars may admit that immaterial entities did appear and that Gods did make themselves felt; they may admit that there are divine *phenomena*. But, they add, such phenomena are not what they seem to be. They are "illusions" and, therefore, *do not count as indicators of reality*. (Democritus long ago, Galileo more recently, and many modern scientists say the same about sensations and feelings.)

The notion of reality behind this account again transcends any set of existing (or even possible) scientific data; it is a metaphysical notion. It also contains a normative component: behavior should center around what is real and must avoid being influenced by illusions. And it has important religious ancestors. Orphism, Pythagoreanism, Gnosticism, Marcionism—all these doctrines subdivide, evaluate ("higher" and "lower" levels of reality), and, accordingly, come combined with ethical demands. Even Parmenides, who tried to *argue* his case, received his basic truth from a Goddess. The religious fervor with which some scientists defend their vision of reality suggests that the connection is rather close. "Biology," said Max Delbrück, the founder of molecular biology, "should be molecular biology entirely. The classical disciplines like zoology and botany should be abolished." Elementary particle physicists occasionally felt the same about chemistry, and behaviorists treat feelings as identified by introspection with even greater contempt. Realists can be tough customers indeed—but there is no reason to be afraid of them.

For what gives them credence is not the power of phenomena but the power of norms evaluating phenomena. We must not be misled by the fact that some phenomena seem to form a coherent whole; if reality were required to produce coherent effects, then shy birds, people who are easily bored, and entities defined by statistical laws would be very unreal indeed. The predicate "real," on the other hand, is only apparently descriptive. Reflecting a preference for forms of coherence that can be managed without too much effort, it contains evaluations, though implicit ones. Now wherever there is a preference there can be, and perhaps should be, a counterpreference. For example, we may emphasize human freedom over easy manageability. This means, of course, that ethics (in the general sense of a discipline that guides our choices between forms of life) affects ontology. It already affected it, in connection with the sciences, but surreptitiously, and without debate. To start the debate we must insert our preferences at precisely those points that seem to support a scientific worldview; we must insert them at the division between what is real and what does not count. And as this division constitutes what is true in science and what is not; *we can say that ethics, having once been a secret measure of scientific truth, can now become its overt judge.*

How will the judge proceed? Being interested in the quality of human lives, he will invert the argument that established the illusionary character of unscientific beliefs and ontologies. Remember the argument. It started from an alleged scientific reality and concluded that people with different surroundings (centered spaces, spiritual entities) lived an illusion. The inversion starts from these lives themselves: Are they really mere dreams, or is there happiness, insight, understanding, affection, gradual growth from childhood via adolescence to maturity and old age? Is there caring for others not only in intention but with good effect?— and so on. Assume all this is established together with even better things than what can be described in a late-bourgeois vocabulary. Then these lives must be respected. We must accept them as human lives in the full sense of the word, i.e., as occurring in a real world, not in a chimera. In other words: *"real" is what plays an important role in the kind of life one wants to live.*

I DON'T KNOW how many of you are prepared to accept this way of looking at things when it is stated as bluntly as I have done just now. Let me therefore repeat that it underlies even those discussions that seem to proceed without an appeal to values and that it plays an important role in politics, common sense, and, in a somewhat hidden way, even in the most advanced parts of science. Two examples to illustrate the situation.

As you may know, Parmenides held that Being does not change and has no parts. This was the first conservation principle of Western science—it asserted the conservation of Being. Parmenides also provided some arguments for his view. They were powerful arguments and quite convincing. Parmenides was, of course, aware of change—but he regarded it as secondary and subjective. Aristotle criticized Parmenides in two ways. He analyzed the *arguments* and tried to show that they were invalid. We may call this a logical criticism. But he also pointed out that Parmenides' *result* would inhibit practical life and political action. This is the kind of criticism I am talking about: a way of life is made the measure of reality. Commenting on Platonists who tried to justify good behavior by reference to a supreme and unchanging Good, Aristotle wrote (*Ethica Nicomachea* 1096b33 ff.—my emphasis):

> Even if there existed a Good that is one and can be predicated generally or that exists separately and in and for itself, it would be clear that such a Good

can neither be produced nor acquired by human beings. *However, it is just such a good we are looking for . . .* one cannot see what use a weaver or a carpenter will have for his own profession from knowing the Good in itself or how somebody will become a better physician or a better general once "he has had a look at the idea of the Good" [apparently an ironical quotation of a formula much used in the Platonic school]. It seems that the physician does not try to find health in itself, but the health of human beings or perhaps even the health of an individual person. For he heals the individual.

"It is just such a good we are looking for," says Aristotle: the needs of a particular form of life determine what is to be regarded as real.

A second example is more technical but perhaps even more interesting. The quantum mechanics of Dirac and von Neumann has a consequence that brings us very close to Parmenides. Using a realistic interpretation of one of its basic principles, the principle of superposition, we have to admit that separate objects are an illusion and that Nirvana is a reality. Physicists did not take this step (except facetiously, when discussing paradoxes like the famous Schrödinger cat). They took it for granted that macroscopic objects exist and have the properties classical physics ascribes to them. They therefore tried various tricks to make quantum mechanics compatible with the existence of such objects. You might say that they simply adapted the theory to "facts" and that no decision was involved. This overlooks that "facts" never prevented a serious realist from retaining a view he found plausible. It is a "fact" that the sun rises and the horizon does not fall—yet this "fact" was regarded as merely apparent by the Copernicans. And scientists are gradually realizing the normative component inherent in all statements of reality. "Everything which is practically real," writes Hans Primas in his path-breaking book *Chemistry, Quantum Mechanics, and Reductionism* ([New York: Springer, 1984], 252), "should appear as objectively real in the theory." Again, a prevalent form of life, the practical attitude of physicists, is taken as a measure of reality.

TO SUM UP: Fang's order of argumentation, which goes from things to norms, can be inverted. It should be inverted by those who want to retain their cultural heritage, who resist being run over by "modern civilization," and who want to change it.

I now come to my last comment on Fang's speech. For Fang a move away from Western civilization and its main ingredient, scientific universality, is a sign of fear. "Those who rejected the idea that science has universal application were in fact doing nothing more than demonstrating their fear of modern culture," he says. Unfortunately we have good reason to be afraid. The claim to universality that started with Western philosophy, and was then taken up by the sciences (and is now being gradually restricted), has led to very undesirable results. Listen first to a quotation from Peter Medawar, a Nobel Prize winner. "As science advances," he says,

> particular facts are comprehended within, and therefore in a sense annihilated by, general statements of steadily increasing explanatory power and compass whereupon the facts need no longer be known explicitly. In all sciences we are being progressively relieved of the burden of singular instances, the tyranny of the particular. ( *The Art of the Soluble* [London: Methuen and Co., 1967], 114)

The quotation sounds a little extreme, but not too dangerous, you will say, provided it is read in the right spirit. Now assume we are talking not about "facts" in general, but about human beings. The "tyranny of the particular" then includes the idiosyncrasies, wishes, and dreams of particular individuals. A scientific approach, says Medawar, "annihilates" them and "relieves" science of their "burden." This may mean that science only deals with some general features of human beings and groups of human beings and leaves the rest to other agencies, but it may also mean that only the selected general features count. I have tried to show that the metaphysical hypothesis that underlies realism takes the second road. I now add that Ceausescu did in practice what the hypothesis (and Medawar) do in theory. He tore down idiosyncratic ancient villages, suppressed idiosyncratic local beliefs, and replaced them by monsters made of concrete and a uniform ideology. You object that it is unfair to mention Ceausescu in the same breath with the universality of science? I don't think it is. The members of the European Community, those standard bearers of Civilization and the Free World, want to bring "backward" regions like Portugal, Greece, and the south of Italy up to their own high level of existence. How do they determine backwardness? By notions such as "gross national product," "life expectancy," "literacy rate," and so on. This is their "reality." "Raising the level of existence" means raising the gross national product and the other indicators. Action follows, as in Fang: monocultures replace local production (example: eucalyptus trees in Portugal), dams are built where people lived before (Greece), and so on. Entire communities are displaced, their ways of life destroyed just as they were in Ceausescu's Romania, they are unhappy, they protest, even revolt—*but this does not count.* It is not as "real" as are the facts projected by an "objective" economic science. Is it not wise to be afraid of such a civilization? And is it not advisable to reverse a way of arguing that encourages the trends I have just described? According to Fang we argue from scientific reality to ethics and human rights. This is a dangerous movement. It does use norms, but hides them behind factual statements; it blunts our choices and imposes laws instead of letting them grow from the lives of those who are supposed to benefit from them. I suggest that we argue the other way around, from the "subjective," "irrational," idiosyncratic kind of life we are in sympathy with, to what is to be regarded as real. The inversion has many advantages. It is in agreement with human rights. It sensitizes us to the fact that Fang's "reality" is the result of a choice and can be modified: we are not stuck with "progress" and "universality." It is plausible because already at the quantum level Being is more ambiguous than the supporters of a realist metaphysics seem to assume. A flea can live in it—but so can a professor and the ontology of a flea will certainly be different from that of the professor. The inversion is not motivated by a contempt for science but by the wish to subject it, this product of relatively free agents, to the judgment of other free agents instead of being frightened by a petrified version of it. Finally, we learn that even a great and committed humanitarian may be inspired by a dangerous philosophy. Good and Evil are close neighbors. We better watch out!

# {10} UNIVERSALS AS TYRANTS

# AND AS MEDIATORS

NIVERSALS PLAY an important role in philosophical and scientific definitions of knowledge. For many writers, to know means to possess an account composed of universals or, more recently, a theory. Some add that given a good theory one can forget about particulars. Thus Peter Medawar writes:

> As science advances, particular facts are comprehended within, and therefore in a sense annihilated by, general statements of steadily increasing explanatory power and compass whereupon the facts need no longer be known explicitly. In all sciences we are being progressively relieved of the burden of singular instances, the tyranny of the particular. (*The Art of the Soluble* [London: Methuen and Co., 1967], 114)

This is in perfect agreement with the attitude of some early Western philosophers.

For Homer, the patterns of geometric and archaic thought and large parts of fourth-century Athenian common sense were not tied to individual cases; they contained generalizations and used universals to give them substance. But these universals did not supersede or constitute particulars, they connected them. We might say that the universals of Plato and Medawar are tyrants which "annihilate" particulars while the universals of their rivals mediate between them while leaving them unchanged. This difference in the use of universals is an important ingredient of the difference between "objectivity" and a "subjective" approach.

Of course, expressions such as "commonsense" or "archaic thought" or "the Homeric worldview" should not be taken too seriously. They sum-

Published in I. C. Jarvie and N. Laor, eds., *Critical Rationalism, Metaphysics, and Science: Essays for Joseph Agassi* (Dordrecht: Kluwer Academic Publishers, 1995), 1:3–14, © 1995 Kluwer Academic Publishers. Reprinted with kind permission from Kluwer Academic Publishers.

marize what scholars inferred from their findings; they cannot be used to define limits of thought and/or sense. We may say "this is how people talked at the time, what they asserted, what they failed to mention"; we go too far when saying "this is what people could think and what was in principle inaccessible to them." Such statements not only exceed the evidence, they clash with it. When adumbrating a moral order that differed from and clashed with the existing social order Achilles raised problems not considered before, he did not raise impossible problems (*Il.* 9.318 ff.). The literary remains of the "Homeric world" may be fairly definite; the "Homeric world" itself was indefinite in the sense that it permitted unusual and (for some scholars) unthinkable moves. It seemed limited because nobody had thought of the moves, and it still seems limited because some scholars have raised this thoughtlessness to a principle.

There is a second reason why common sense and related ways of thinking seem so clearly circumscribed: their philosophical critics presented them in this way. Trying to create a new culture they gave simple, concise, and easily assailable definitions of the old. The victims accepted the definitions and further stabilized the thing defined: the "common sense" attacked and defended by philosophers and "innocent" common sense are not the same thing.[1] Change itself added to the freezing of worldviews. An agent effecting change reveals the ambiguity of the status quo. He uses the ambiguity to introduce new elements which he then clarifies by confronting them with a well-defined past. Having been constrained in this manner a way of life may indeed start looking like a "system of thought." It is such a "system," such an artifact, and not its unreflected source which I am going to compare with the ways of the philosophers.

In his essay *Die Ausdrücke für den Begriff des Wissens in der vorplatonischen Philosophie* (Berlin: Weidmannsche Buchhandlung, 1924), Bruno Snell examines how older Greek authors described the process of obtaining and justifying information. Snell found not just one expression, or an expression with modifications, he found an entire spectrum of situations and ways of behaving. Our own language is similar in this respect. A reporter can substantiate a story by saying: "I have seen it," or "I

---

1. This is one of the main themes of E. Husserl, *The Crisis of European Sciences and Transcendental Phenomenology,* trans. D. Carr (Evanston: Northwestern University Press, 1970). Husserl concentrates on post-Galilean developments. Trouble started in antiquity.

have heard about it," or "I was there, I even participated," or "I know what it means to be poor; for years I had hardly anything to eat, I slept in parks or in railway stations, occasionally friends put me up—but their friendship did not go very far; they soon lost respect—well, money seems to play an important role in such matters"—and so on. Knowledge so described is as complex as are the idiosyncrasies of those who claim to know and the surroundings in which they act, suffer, feel, speak, learn, and explain what they have learned. And justifying something does not mean relating it to an abstract entity such as "experience," "experiment," a principle of reason or an ingredient of Husserl's "Lebenswelt"; it means telling a story that includes a personal guarantee. (It seems that Protagoras's "man is the measure of all things" was meant in precisely this way.)

In ancient Greece the diversity of personal interests and types of information found expression in a comparable diversity of the so-called arts *(technai)*: people with similar aims and talents united, they created special procedures and languages and, on that basis, special products such as the epic, drama, statues, buildings, weapons, metals, treatment of the sick, geometric theorems, and many other things. This separation of the arts and, thus, of different types of knowledge was in part new, in part a very ancient matter. Most likely it arose with agriculture and the corresponding settled ways of life; individual hunters and gatherers had known all the tricks that were needed for survival. But while the arts, taken together, preserved at least part of the diversity of knowledge, the information *inside* each *techne* became increasingly standardized: you do not find personal features in the Babylonian-Assyrian word lists (which may be regarded as a first very elementary form of science).[2] Not the scribe, his profession guarantees the correctness of the information.

For the early Greek philosophers and then especially for Plato this already very tame diversity was still much too wild, or, to use a modern expression, too "subjective." In his dialogues Plato inquires into the nature of courage, piety, virtue, knowledge, etc. The first answers "Socrates" obtains are lists. Thus Theaetetus, being asked for the nature of knowledge (146c3 ff.), provides a list of the professions which in his opinion lead to

---

2. W. von Soden, *Leistung und Grenze sumerischer und babylonischer Wissenschaft* (Darmstadt: Wissenschaftliche Buchgesellschaft, 1965).

knowledge: geometry, astronomy, the arts of the cobbler "and of all other artisans." The answer corresponds to the situation of the time; the domain of knowledge was indeed divided into sections and a list, therefore, is a very good account. Socrates is not satisfied: "You are very generous and liberal, my dear friend! Being asked for one thing you give me many, and diverse things when I am asking for something simple." What is behind this objection? The assumption that knowledge is simpler than Theaetetus seems to suggest. How is the assumption supported? By the form of the question. Socrates asked: "What do you think that *knowledge* is?"—i.e., what a single thing is. The unitarian form implies a unity of subject matter only if the word "knowledge" does not simply collect particulars, one by one, but represents a property that is common to all of them. But that is the problem in question! Even worse; in preparing his inquiry Socrates uses three different epistemic terms: *episte'me* (and the corresponding verb), *sophi'a* (and two further forms belonging to the same root), and *mantha'nein*. Thus the use of *episte'me* in the main question seems purely accidental, there is no foundation for it, and the occurrence of one word is no objection against a list. Are there better reasons?

In 147d4 ff. Theaetetus describes a mathematical discovery. On this occasion he also speaks about incommensurability. The passage has been discussed at length, though with widely differing results and different translations of even very simple words. As the point I want to make is independent of these differences I shall not dwell on them except perhaps for remarking that it makes no sense trying to obtain precise information from a series of rather ambiguous statements.[3] Incommensurability appears as an old acquaintance; it is not the problem. There is not a trace here of the "ancient foundational crisis" which some authors connect with the discovery of lengths incommensurable to the unit. The problem

3. Historians have analyzed the passage as if it were a professional paper on the history of mathematical procedures. But quite apart from the fact that Plato explicitly rejected the scholarly style (*Phaedrus* 274c5 ff. and other places), he was not concerned with doing history, but with giving an example of a definition different from a list. The example is straightforward and this quite independent of the question whether it represents a major mathematical advance or a boyish triviality (an important issue of the debate, with van der Waerden and Szabo taking opposite positions). For a survey of some of the problems raised by the passage, references, and a judicious approach cf. M. F. Burnyeat, "The Philosophical Sense of Theaetetus' Mathematics," *Isis* 69 (1978): 489 ff.

is whether numbers which are incommensurable with the unit have common properties. This is what Theaetetus says (147d4 ff., paraphrase after John McDowell):

> Theodorus here was drawing diagrams to show us something about squares—namely that a square of three square feet and one of five square feet aren't commensurable, in respect to length of side, with a square of one square foot; and so on, selecting each case individually, up to seventeen feet. At that point he somehow came to an end. Well, since the squares seemed to be unlimited in number, it occurred to us to do something along these lines: to try to collect them under one term by which we could refer to them all.

Theodorus showed the irrationality of the square roots of three, five, and so on until seventeen. He showed this individually, for each single number; in other words, he produced a list of all irrational square roots of integers, from that of three to that of seventeen. Theaetetus (and his friend, the younger Socrates, who participated in the demonstration) then tried to characterize irrationals in a different manner, not via a list, but by means of a procedure that would unite them under a single concept. Theaetetus describes the procedure as follows:

We divided all numbers into two sorts. If a number can be obtained by multiplying some number by itself, we compared it to what's square in shape, and called it square or equal sided.

SOCRATES: Good.

THEAETETUS: But if a number comes in between—these include three and five, and in fact any number which can't be obtained by multiplying a number by itself, but is obtained by multiplying a larger number by a smaller or a smaller by a larger, so that the sides containing it are always longer or shorter—we compared it to an oblong shape, and called it an oblong number.

SOCRATES: Splendid. But what next?

THEAETETUS: We defined all the lines that square off equal-sided numbers on plane surfaces as lengths, and all the lines that square off oblong numbers as powers, since they aren't commensurable with the first sort in length, but only in respect of the plane figures which they have the power to form. And there's another point like this one in the case of solids.

Theaetetus introduces four new terms: length, power, square number, oblong number. Length is defined as the side of a square whose area is a square number, power as the side of a square whose area is an oblong number. The definitions led to the theorem (which is not presented, but implied) that *irrational numbers are powers.* Instead of a *list* we have a *property* (created by a *definition*) shared by all irrational numbers and permitting us to "collect them under one term."

The discovery that long and open lists or stores can be changed into short and closed theorems by means of a new entity, a "concept," and that the schematisms that lead to such a concentrated way of speaking, viz., concepts, can be constructed and managed with the help of special sequences of words, so-called definitions, was recent and represented an *extension* of common sense. The extension leads to a *unification* only if the new thing, power in our example, stands in a closer relation to the things to be collected (the numbers) than the old concept of irrationality—for otherwise we would again have to prove irrationality for every single number, only this time under a different name. Now the algorithm of decomposing a number into primes is simpler and more general than the geometric version of the subtraction algorithm *(anthyphairesis),* which Theodorus seems to have used (he was a geometer and was "drawing diagrams," according to the story).[4] The question is if such simple algorithms can also be found in the case of knowledge.

The question has two sides: (a) *should* knowledge *be changed* in such a way that its presence can be checked by a simple algorithm and (b) *was* the *existing* knowledge sufficiently simple to permit checking by such an algorithm. At the time of Plato and Socrates, when professions had become well established, the first question could be classified as a political question: which kind of knowledge do we want in our community? (Plato turned the question around: which kind of community fits the type of information that counts as knowledge.) The second question is a historical question. The answer is that fundamental concepts such as the concept of knowledge or the concept of virtue had already undergone a decisive change. Kurt von Fritz and Bruno Snell describe how the rich spectra of epistemic and value expressions which I mentioned toward the

4. Cf. B. L. van der Waerden, *Science Awakening* (Groningen: Noordhoff, 1954), chap. 5. Cf., however, D. H. Fowler, *The Mathematics of Plato's Academy* (Oxford: Clarendon Press, 1987), where completely new light is thrown on the problem.

beginning slowly shrank, how the elements of the spectra lost their relation to individual events and contexts, and how concepts became impoverished, "empty and one-sided" (von Fritz). Similar developments occurred in the domains of law and economics, and here especially after coins, which were in themselves worthless, had replaced barter and the exchange of gifts. Like the method of Theaetetus, such coins assemble objects with different individual properties under a single abstract concept, their "monetary value." Question (a) now becomes very important. For the abstract monetary "value" of an object was not something that had existed at the time of barter but had been discovered only recently; it was part of a process that had destroyed old social ties and replaced them with different and more abstract connections.

In a similar way Plato's simplifications were not intellectual games; they had practical consequences which, as I said, were paralleled by the thought, the speeches, and the actions of the time. Should these developments be supported or should one try to hinder them? After all, the "older ways" still existed. Their influence was enormous and reached even into the closed world of the Platonic dialogues: "But I am unable, Socrates, to give you a similar answer about knowledge," says Theaetetus at the end of his mathematical presentation and in this way expresses the great difference that exists between a mathematical theorem and other equally important types of information.

Plato's discussion of the nature of virtue has similar features. In *Meno* 71d Socrates asks for a definition of virtue. Meno answers (71e):

> It is not difficult to reply to your question, Socrates. First, if you will, the virtue of a man; it is easy to see that the virtue of a man is that he is capable of managing the affairs of the state, that in doing so he takes care of his friends and hurts his enemies [an old Greek saying] while also taking care that bad things do not catch up with him. If you want the virtue of a woman, that, too, is easily described; she must manage the household well, keep everything in the house in good shape, and obey her husband. Still another virtue is the virtue of a child, of a boy and a girl, or of an old person, be he a free man or a slave. And there are many other virtues.

Like the first answer of Theaetetus the answer of Meno is factually adequate. The inquiry concerns matters which play an important role in social life. A society that is not subjected to tyrannical laws has many customs and types of information; it contains groups, arts, crafts, sciences with different professional commitments; actions in such a society are fitting or inappropriate, good or bad, acceptable or out of touch depending on the circumstances in which they are performed; these change with time, they are not always codified or written down, which means that behavior cannot be ruled by an algorithm of general concepts—the individuals themselves must decide or, to put it differently, they must treat their surroundings in an inventive way, producing new responses to new phenomena. The answers given by Theaetetus and Meno take this complexity into account. They provide *examples* and thereby invite us to consider more than one type of action; they use *a variety of* examples and thereby prevent a uniform interpretation of the material; and they are *imprecise,* for matters are in flux and cannot be nailed down. In short, the answers, though apparently naïve and childish when viewed through standardized conceptual spectacles, are realistic (and tolerant). And, what is most important, they are not sudden inventions dreamed up to deal with an absurd question; they reflect widely accepted opinions, slogans, worldviews. The Sophists, too, used lists, stories, loose paradigms to keep their listeners in touch with the abundance of life in a city-state. The tragedians—Sophocles *(Antigone)* and Euripides especially—showed that a "systematic" treatment of the virtues eliminates the element of conflict which in their opinion characterizes all human existence. Greek morality at the time of Plato was a morality of instances and examples, not a morality ruled by abstract properties.[5] The Homeric epics, which continued to play an important role in education,[6] introduced the virtues through stories which illustrated both their nature and their limits. Identification (of the listener or reader with the story) was an important part of the illustration: not the epos (or, later, drama) decided about the limits; it was the listener or the reader who, having been

5. For the Sophists, their ancestors, and their opponents, cf. Fritz Wehrli, *Hauptrichtungen des griechischen Denkens* (Stuttgart and Zurich: Artemis, 1964). Virtue is discussed in Martha C. Nussbaum, *The Fragility of Goodness* (Cambridge: Cambridge University Press, 1986). Cf. also K. J. Dover, *Greek Popular Morality at the Time of Plato and Aristotle* (Berkeley and Los Angeles: University of California Press, 1974). George Steiner, *Death of Tragedy* (New York: Knopf, 1961), examines why after Shakespeare and Racine tragedy seemed *difficult to achieve.* It was *opposed* in antiquity by those who, like Plato, looked for a consistent definition of virtue. Cf. *Euthyphro* 7b ff.

6. T. B. L. Webster, *Athenian Culture and Society* (Berkeley and Los Angeles: University of California Press, 1973), chap. 4.

drawn into the action, noticed when courage became savage rage or wisdom slyness and drew the limits accordingly. The justification was personal, not abstract. The wise men of *Il.* 18.503 show how this personal element functioned in ancient law.

All this leaves Socrates unmoved. With the typical (genuine or pretended) blindness of the intellectual he answers:

> I seem to have been very lucky, my dear Meno; I looked for a single virtue and I found an entire swarm of them;

and he points out that bees are unified not by enumeration, but by means of a few biological properties and that health and sickness are the same for man and woman. Still, Meno hesitates:

> Somehow it seems to me that the case [of the virtues] differs from the examples you just gave.

Thus Plato introduces an obstacle at precisely those places and with respect to precisely those concepts which the epic, tragedy, comedy, and fourth-century Athenian common sense had explained by stories, lists, examples, and not by giving definitions—this is the only indication that Meno's answer is more than childish confusion. It is easy to understand why Theaetetus and Meno hesitate. Numbers, perhaps even bees, may be simple things—the same for Greeks, barbarians, Athenians, Persians, Spartans; it may be possible to characterize them with the help of general definitions.[7] But customs, virtues, information change from one city to the next and even more so from one nation to another. Herodotus had given a colorful account of this multiplicity. Searching here for a common property seems hopeless—and yet Socrates proceeds as if it were the most natural thing to do. We can assume that the property sought—if it exists—will be either empty or tyrannical. A glance at the work of early philosophers confirms this suspicion. They do not enrich existing concepts, they void them of content, make them crude, and increase their influence by turning crudeness into a measure of truth. They clearly distinguish between "the many" who keep older traditions alive (and cannot be taken seriously) and "the few" who know truth and are high

---

7. A. Seidenberg, B. L. van der Waerden, and others have suggested that the widespread use of simple laws of arithmetic and geometry in ancient times was due to inventions in one place and diffusion to others. It was the result of a historical accident and not of Platonic constraints (the invention itself is explained by the requirements of ritual).

above the multitude:[8] the "Rise of Rationalism in the West" shows the same kind of contempt toward nonbelievers that accompanied the rise of modern science.

On the other hand, there is no sign that traditions based on imprecise and open concepts are losing in strength. They survived the onslaught of rationalists in antiquity. Occasionally they treated the philosophers with considerable sarcasm.[9] On other occasions they were simply pushed aside[10]—but they always returned, in religion (personal religion vs. theology), medicine (medical practice vs. theory), biology (the work of naturalists down to Konrad Lorenz vs. the work of theoreticians, molecular biologists especially), mathematics (constructionism vs. Platonism). A look at "high theory" shows that the formulae expressing it, though perhaps intended as outlines of universal features of the world, often merely serve as summaries of approximations obeying a variety of symmetry principles—they are mediators, not tyrants as some so-called "realists" assume. The tendencies exist side by side with each other and with the arts whose popularity at any rate by far outweighs that of the sciences (film, videos, rock music, heavy metal, etc.). The situation certainly is complex—but what is the balance?

I don't think that there exists a satisfactory answer to this question. I personally don't have the knowledge (I don't think anyone has) and even a well-informed survey could at most produce impressions, not the comparative judgment that is needed. Besides, we must not forget that history looks different and moves in different directions, depending on the background of those approaching it. Yet it seems to me that the process called "Western civilization," which grew in our midst, then spread all over the world and filled it with knowledge, stench, weapons, and monotony, is not the mixture of abstractness and subjectivity some newspa-

8. Details in Hans-Dieter Voigtlaender, *Der Philosoph und die Vielen* (Wiesbaden: Franz Steiner Verlag, 1980).

9. An example is the criticism of abstract theories of well-being by the author of *Ancient Medicine* (secs. 15 and 20), in *De prisca medicina*, chap. 15 (*Corpus Medicorum Graecorum* 1, 1). Cf. also "Realism," essay 4 in part II of the present volume.

10. The reasons were anything but compelling. Thus early in the eighteenth century physicians and biologists, assuming that Newton had succeeded by uniting everything under a single principle, "gravity," tried to build analogously unified systems. But Newton had not used a single principle, he had used many; his success was not due to a coherent procedure, but to numerous ad hoc adaptations, and, besides, there were still many unsolved problems. A chimera, not a real thing, rattled the bystanders.

pers and university programs make it out to be. It is a process in which impersonal factors, accidents, greed, cowardice, and strong destructive forces have long played a decisive role. Scientists, not poets or local housewives, are called upon in ecological and technological projects, in plans about the rebuilding of cities and counties; scientists, not peasants or small business people, get funds (they still do—even in 1991) to secure peace by militarizing the sky. Old ways of living are being destroyed and replaced by factories, highways, and monocultures which turn the science-based principles of experts (economists, agronomists, engineers, etc.) into tyrants without paying attention to local wishes and values. They do not mediate between those wishes, they suppress them. All this is being done not just by maniacs like Ceausescu (who made his own contribution to the trend) but by the "free world," in Spain, Greece, Portugal, with EEC, World Bank, and international aid money as a motor. The intention is to bring these countries "up to the level" of the rest of Europe. But "coming up to the level of the rest of Europe" (Italy, for example, or Germany) does not mean that individuals are now going to be happier and are going to lead a more fulfilling life—it means an increase of abstract entities such as the "gross national product," the "growth rate," and so on. Elsewhere I have shown how the phenomenon started in antiquity. Parmenides removed personal concerns, Protagoras reintroduced them—though in a language already modified by the abstract thought of his predecessors. Like them he is said to have described humans in abstract terms, his fault (in the eyes of his opponents) being that he used the wrong terms and the wrong entities—sense data instead of ideas. But sense data have little to do with human experience—they are as abstract and, indeed, as inaccessible as pure Being. Originally, notions involving people were introduced by examples, in the course of a complex but never entirely transparent practice (cf. Meno's reply to the question "what is virtue?"). Now the critics of tyrannical universals want to defend "the subject": they have adopted what they claim to reject, the categories and the methods of their opponents. (Relativism, too, insofar as it is not simply a call to tolerance opposes objectivism *within philosophy;* it has lost its connection with the worldviews it tries to defend.)

But can we live without universals? Is it possible to increase our knowledge and yet to preserve its looseness? And does the suppression of what is genuinely subjective not already start in personal relations and then even more so in the realm of politics, which cannot exist without something that is shared by all? "Speaking with understanding they must hold fast to what is shared by all, as a city holds to its laws, and even more firmly," writes Heraclitus (fragment 114, trans. Charles H. Kahn). Agreed. But all depends on how "what is shared" is reached and how it rules once accepted.

In a dictatorship "what is shared" is imposed; it rules because its subjects are not allowed to act otherwise. "What is shared" rules also in a democracy but not because its citizens either cannot think and act in any other way or because they have been forbidden to think and act in any other way; "what is shared" rules because the citizens have decided to orient their public actions (*not* all their actions) temporarily (*not* forever) around a simple program (*not* around a "rational foundation" or a "humanitarian ideal"—though either can play a role in the choice of the program). The program is conceived by individuals, it rests on their idiosyncrasies, and it disappears when they do.[11] The procedure can be extended to the interaction of cultures. Cultural collaboration presupposes that there are things "that are shared." Between whom? Between the interacting parties. In what area? In the area of interest. Who decides about the common elements? The parties themselves, if possible in collaboration with those powers that are affected by the collaboration. According to what principles? According to the principles the parties either bring to or invent during the negotiations. (As more recent historians have shown this is also the way in which clashes between different scientific schools are negotiated.) Should relations between humans not be guided by universal principles? Yes, if such principles happen to have universal appeal; no, if their actual power among the negotiating parties is limited. About 2000 B.C. simple mathematical ideas and constructions were known in China, India, through the Near East up to the British Isles. They were known because they had been invented, had spread, and had been accepted. Naturally, they played a role in any cultural exchange

11. Interestingly enough the generalities that emerge from scientific practice (as opposed to the global generalities the older philosophers used when talking about the sciences) are also contingent features of this practice which linger for a while, then change and eventually disappear (example: the disappearance, after the arrival of large-scale experimental enterprises, of the demand for repeatable experiments). They are not transhistorical agencies. Cf. my "The End of Epistemology?" in *Physics, Philosophy, and Psychoanalysis: Essays in Honor of Adolf Grunbaum,* ed. R. S. Cohen and Larry Laudan (Dordrecht: Kluwer, 1993).

involving numbers. Today a rather concrete idea of freedom and humanity influences actions in Western and Eastern Europe, and, though as yet unsuccessfully, in the Far East; it guides revolutionaries, business enterprises, and to some extent even the actions of more conservative bodies. This is very much to be welcomed. What is not so welcome is the attempt to again tie a process that is in flux to transhistorical agencies or to freeze the principles that push it along; what is not to be welcomed is the attempt to turn words and concepts that mediate between people into Platonic monsters that rebuild them in their image. (Paradoxically, intellectual fighters for freedom and enlightenment at all ages—with very few exceptions—tried to do just that.) What is not to be welcomed is a universality that is enforced, either by education, or by power play, or by "development," this most subtle form of conquest. But is not science universally true in the sense I am trying to criticize and does it not show that Platonic universality has come to stay? My answer is the same as before: assume that science is universally accepted (which it is not, and cannot be, for "science" as a single uniform entity is a metaphysical monster, not a historical fact)—then this would be a historical accident, not proof of the adequacy of Platonic universals—and one might try to change it.

# {11} INTELLECTUALS AND THE FACTS OF LIFE

ANCIENT NAVIGATORS, jewelers, farmers, architects, merchants, blacksmiths, shipbuilders, physicians, chroniclers were familiar with a great variety of materials, plants, animals; they could identify and alleviate bodily and mental afflictions; they traveled across national boundaries and assimilated foreign ideas and techniques. Archaeological discovery shows how much was known, for example, about the properties of metals, their compounds and alloys, and how skillfully this knowledge was used. An enormous amount of information resided in the customs, the industries, and the common sense of the time.

Most Greeks took this information for granted. Not all of them were impressed by it. Aiming at something more profound, some early thinkers started the work of knowledge all over again, this time without details but with a maximum of generality. They were religious reformers, for they heaped scorn on the gods of tradition and replaced them with monsters. They were philosophers, for they preferred words to things, speculation to experience, principles to rules of thumb; and they did not mind when their ideas conflicted with tradition or with phenomena of the most obvious kind. They were also rather conceited. Being convinced of the superiority of their efforts, they upheld their side of the conflict: the phenomena were spurious and tradition was worthless. Thus Parmenides claimed that the world was one, that change and subdivision did not exist, and that the lives of human beings, which contained both, were a chimera.

The doctrine has survived until today. Some nineteenth-century scientists and their followers in the twentieth century posit a "real" world, without colors, smells, and sounds, and with a minimum of change; all

Published in *Common Knowledge* 2, no. 3 (winter 1993): 6–9. Reprinted by kind permission of Oxford University Press.

that happens in that world is that certain configurations move reversibly from one moment to another. In a relativistic universe, even those events are laid out in advance: the world "simply *is,* it does not *happen.* Only to the gaze of my consciousness, crawling upward along the lifeline of my body, does a section of this world come to life as a fleeting image in space which continually changes in time."[1] "For us who are convinced physicists," wrote Einstein, "the distinction between past, present, and future has no other meaning than that of an illusion, though a tenacious one."[2] The trouble is that Einstein was also an empiricist: science must be checked by experiments which are spatiotemporal processes and, therefore, "illusions." But how can illusions guide us toward reality?

The first Western philosopher to address the problem in full generality was Aristotle. Considering the conflict between abstract principles and common sense (artisan-practice), he opted for the latter and modified philosophy accordingly. In his critique of Parmenides, he did not emphasize the *fact* of change—after all, this fact was well known, though Parmenides had declared it to be a chimera. Aristotle instead argued that *change played an important role in human life* and that it would therefore be "next door to lunacy" (*De generatione et corruptione* 324a18 ff.) to regard it as unreal: human existence or, more specifically, *a well-established practice,* not theory, was his measure of reality.

Cultural exchange is a well-established practice. For millennia, cultures exchanged ideas, technological achievements, art forms, luxuries, foodstuffs, deities, prostitutes; their writers (if there were any) described each other's peculiarities, their generals paid attention to foreign methods of warfare, rumor entertained those not immediately involved in the process. Naturally, there were mistakes, misunderstandings, and distortions. Intolerance was never far away. On the other hand, we have many accurate observations, some of them revealing an astonishing insight into originally unfamiliar customs and beliefs. Insight did not always allay intolerance; after all, even brothers and sisters may have become enemies. But as in the case of nature, it seemed possible to discover errors and to acquire more adequate views. And as in the case of nature, there arose thinkers who analyzed this belief and declared it to be an illusion. This, if I understand them correctly, is the problem that bothers

1. Hermann Weyl, *Philosophy of Mathematics and Natural Science* (Princeton: Princeton University Press, 1949), 116.

2. Einstein, *Correspondance avec Michèle Besso* (Paris: Hermann, 1979), 312, 292.

Messrs. Appiah, Frank, and Katz in their Call for Papers (*Common Knowledge* 1, no. 2, p. 1). They are aware of the work of "explorers, lexicographers, traders"; they have read what happened "in philosophy since the Forties"; and they envisage a form of intercultural understanding that, being guided by "the philosophical principles of incommensurability and indeterminacy of translation," is "not theoretically naïve." What they fail to consider is that it may be the principles that are naïve.

Take incommensurability. Following Kuhn (who, incidentally, is much more subtle than many of his imitators), some have asserted that the transition from a comprehensive physical theory to its historical successor involves an act of conversion and that the converted no longer understand the older faith. That is not what we find when we look at history. The transition from classical physics, with its objective space-time frame, to the quantum theory, with its attendant subjectivities, certainly was one of the most radical transformations in the history of science. Yet every stage of the transformation was discussed. There were clear problems; they worried both the radicals and the conservatives. Many people suggested solutions. These solutions, too, were understood by the contending parties, though not everybody liked them or regarded them as important. The final clash between the new philosophy and its classical predecessor found its most dramatic expression in the debate between Bohr and Einstein. Did Bohr and Einstein talk past each other? No. Einstein raised an objection; Bohr was mortified, thought intensely, found an answer, told Einstein, and Einstein accepted the answer. Einstein raised another objection; Bohr was again mortified, thought intensely—and so on. Looking at such details, we realize that the conversion philosophy simply does not make sense. I suggest that the notion of "hermeneutic frontiers" has similar shortcomings, and for exactly the same reasons.

Unfortunately, this is not the end of the story. Parmenides made assertions that prima facie were "next door to lunacy." Yet the assertions (or a more general view that may have led to them) had effects—in mathematics (axioms instead of constructions and intuitive symmetry considerations), physics (conservation principles, belief in inexorable laws), cosmology (steady-state cosmologies; sharp separation between human perception and physical reality)—and so on. Philosophical theories that throw doubt on the possibility of intercultural understanding are equally implausible. Yet they, too (or the motivations behind them), have effects such as the idea that only an Asian actor can play the part of an Asian, or that only an African American can write the history of slavery in the United States, or that only a disabled person can understand what it means to be disabled.

Now I can image a world that has precisely these features. It is not our world (disabled persons are being helped—also emotionally—by researchers without any disability, and African American historians have learned a lot from their white colleagues), and it is not a world I could endure. Do I have a choice? Yes, I do. Our surroundings, the entire physical universe included (see my column in *Common Knowledge* 1, no. 3), are not simply given. They respond to our actions and ideas. Theories and principles must therefore be used with care. Most of them exclude specifics and personal matters; speaking bluntly (though not untruthfully), we can say that they are superficial and inhumane. Being applied with skill and determination, they turn genuine understanding, which is a live process involving more than one party and depending on all of them, into a scholarly artifact. Messrs. Appiah, Frank, and Katz agree. The studies they envisage "would presuppose . . . that the context in which two cultures may be understood together will be *constructed* . . . by a scholar." "It will be constructed"—this means that it will not emerge from personal contact; and it will be constructed "by a scholar"—this means that the victims have no say in the matter. "Ilongot introduced me to their world," writes Michelle Z. Rosaldo in *Knowledge and Passion,* making precisely this point; "but they could not teach me to translate the categories of their experience into the descriptive and theoretical discourse of Western Social Science."[3] Thus an approach that was "developed to protect non-Western cultures from intellectual invasion" (Appiah et al.) ends up by giving us a caricature. What happened?

I think that the explanation is rather simple. Understanding cannot exist without contact. Contact changes the parties concerned. Those who are unwilling to change ("scholars" will hardly drop "the descriptive and theoretical discourse of Western Social Science"), and who, in addition, are afraid of changing others ("protect non-Western cultures," etc.), will therefore find themselves in an artificial world that is perfectly described by the "philosophical principles of incommensurability and indeterminacy of translation."

3. Rosaldo, *Knowledge and Passion* (Cambridge: Cambridge University Press, 1980), xiv.

# {12} CONCERNING AN APPEAL

# FOR PHILOSOPHY

I HAVE BEFORE ME a document that is signed by philosophers (e.g., Gadamer, Derrida, Ricoeur, Rorty, Putnam), scientists (e.g., Sciama, Prigogine), and administrators (e.g., the president of the European Parliament). It appeals to "all parliaments and governments of the world to introduce, support, and underwrite with full force the study of philosophy and its history and the related history of the natural humane sciences—from the intellectual treasures of the Greeks and the great Oriental cultures to the present." Such a study, we read, "is the ineradicable presupposition for every genuine encounter between peoples and cultures, for the creation of new categories to overcome existing contradictions and to be able to direct humanity on the path of goodness." "In this hour of movement and history," the document concludes, "we need cultural and civic consciousness. We need philosophy."

The appeal calls philosophy "an eternally effective *elixir* of life." It is the very opposite. Philosophy is not a single Good Thing that is bound to enrich human existence; it is a witches' brew, containing some rather deadly ingredients. Numerous assaults on life, liberty, and happiness have had a strong philosophical backing. The rise of philosophy in the West or "the long-lasting battle between philosophy and poetry" (Plato, *Republic* 607b) is the oldest and most influential assault of this kind.

Most intellectuals believe that the early philosophers showered us with "intellectual treasures." They overlook that the "treasures" were not *added* to the already existing ways of living; they were supposed to *replace* them. The trouble is that the "treasures" lack important ingredients of a rewarding human life, and that, in the search for objectivity and stability, their inventors had the tendency to push such ingredients aside.

Published in *Common Knowledge* 3, no. 3 (winter 1994): 10–13. Reprinted by kind permission of Oxford University Press.

Compared with poetry and common sense, philosophical discourse is barren—and insensitive. It frowns upon the emotional bonds and the changes that keep humans going, which means that philosophers have destroyed what they have found, much in the way that the standard-bearers of Western civilization have destroyed indigenous cultures and ways of life and replaced them with their own idiosyncratic "treasures."

According to Parmenides, human beings, or "the many" as he calls them somewhat contemptuously, "drift along, deaf as well as blind, disturbed and undecided," guided by "habit based on much experience" (Diels-Kranz, *Die Fragmente der Vorsokratiker* [Zurich: Weidmann, 1985], fragments B6.7, B7.3, and B6.6 ff.). Their fears and joys, their political actions, the affection they have for their friends and children, the attempts they make to improve their own lives and the lives of others, and their views about the nature of such improvements are chimeras. According to Plato, most traditional instruments for the presentation and examination of knowledge—the epic, tragedy, lyrical poetry, the anecdote, the scientific treatise (including the many data collected in the Hippocratic writings)—are either deficient or deceptive: they must be changed. Medical practice, for example, must be guided by theory, which can overrule the obtained knowledge of practicing physicians. The arts have no place in an orderly society (*Republic*, bk. 10). But let us see how some of the changes intended by the likes of Plato would have affected particular areas of Greek society; for example, morals.

The early Greeks knew four basic virtues: courage, justice, piety, and wisdom. According to Protagoras (see Plato, *Protagoras* 327e ff.), one learns these virtues as one learns Greek, without an appeal to specialists, without teachers, simply by growing up in a community that practices them. Education is community centered. The Homeric epics reflect this situation. They do not define, they use examples, including cases that show, without explicitly saying so, under what circumstances a virtue turns into a vice. Diomedes is courageous (*Il.* 5.114 ff.); his courage occasionally gets out of hand; in these cases he behaves like a madman (330 ff., 434 ff.). Not the author, but the listener (or, in our days, the reader) makes the judgment and derives the limits of courage from it. Wisdom receives a similar treatment. Odysseus often acts in a wise and well-balanced manner. He is asked to speak to temperamental stars like Achilles; he is sent on difficult missions. But Odysseus's wisdom, too, occasionally changes face and turns into slyness and deception (*Il.*

23.726 ff.). Such instances show what courage and wisdom are; they make us aware of the virtues' complexity; they encourage us to further enrich the virtues by extension and modification, either in our imagination or by acting courageously and wisely in new situations; they do not nail the virtues down, they leave it to us to preserve or alter them: examples teach while encouraging spontaneity.

Concepts so introduced are not Platonic entities, they are not "objective," i.e., separated from things and traditions. Concepts of this kind are on the same level as color, swiftness, beauty of motion, expertise in the handling of weapons and words. They are affected by the circumstances in which they arise, by dreams, emotions, wishes; they are not subjected to rigid rules. The best way of explaining such concepts is to immerse the questioner in the practice that contains the concepts and to ask him or her to act. The second best way is to give an open list of particular instances. And lists played indeed a most important role in the growth of Near Eastern (Sumerian, Babylonian, Assyrian, early Greek) knowledge.

Socrates proceeds differently. Responding to Protagoras's assertion that one learns the virtues as one learns Greek, without a teacher, by simply living within a certain community (tradition), he asks (*Protagoras* 329a ff.) how the virtues are related to each other and to virtue itself. Is courage part of virtue, and if so, then what does "being part of" mean here? Is it opposed to cowardice and, if it is, how is it related to wisdom? Does the fact that courage and wisdom are both "part of" virtue imply that they are the same and, if they are not, what is the relation between them? These are special questions adumbrating new and highly technical concepts with no obvious relation to the day-to-day problems of ordinary people. Socrates is not only interested in such concepts; he insinuates that true virtue (or true knowledge, or true justice, or true piety) are achieved only when, by a kind of trickle-down process, the new concepts have replaced tradition. Being clear and rigorous, these concepts lack the elasticity, the many-sidedness, the emotional aspects, and the familiarity of the traditional concepts of virtue. Having been introduced to solve technical problems, the philosophical concepts also lack the usefulness of their predecessors. The replacement, therefore, will not enrich the lives of the common people, it will make them barren, inhumane; it will also make them unfree, for it will subject them to the unsympathetic, because "objective," judgment of experts. So much for the "treasures" created by the rise of Western philosophy.

The recent appeal to "all parliaments and governments of the world" (etc.) has similar drawbacks. It envisages "the creation of new categories to overcome existing contradictions and to be able to direct humanity on the path of goodness." This may sound reasonable to the ears of intellectuals accustomed to replacing real-world relations by relations between conceptual artifacts. But note what is implied. The categories are not being *offered* to "humanity"; "humanity" is not invited to *consider*, perhaps to change or even to reject them; the categories are to *"direct"* humanity as a policeman directs traffic.

Now it is clear that "categories," taken by themselves, cannot "direct" anything unless they have power, i.e., unless they are imposed by an influential worldly agency. To obtain the power, Plato consorted with tyrants. The appeal asks "all parliaments and governments of the world to introduce, support, and underwrite with full force the study of philosophy"—i.e., education or, considering the nature of government-directed education, brainwashing is supposed to do the trick. What will be the effect of an education based on the "new categories"?

The categories are supposed to "overcome existing contradictions"— the many ways in which people have arranged their lives will be trimmed to fit the categories. Not case-by-case negotiations between the members of various societies, which might preserve some of the richness of world culture, but an overall system, concocted by academic specialists and supported "with full force" by parliaments and governments, is supposed to eliminate the conflict. That is the colonial spirit again, but concealed, as some earlier forms of colonialism were, by treacly humanitarian phrases.

My second criticism is that the appeal is self-serving (philosophers and scientists want their subjects to have greater power) and abounds in big words and empty generalities. The real problems of our time are not even touched upon. What are these problems? They are war, violence, hunger, disease, and environmental disasters. The warring parties have found a wonderful instrument for "overcoming existing contradictions"—ethnic cleansing. The appeal has nothing to say about these atrocities; in a way it even supports them by its proposed method of conceptual and/or cultural cleansing. The philosophers and scientists who signed it would have done better to issue a strongly worded condemnation of the crimes and the murders that occur in our midst, together with an appeal to all governments to interfere and stop the killing, by military force, if necessary. Such a condemnation and such an appeal would have been understood, it would have shown that philosophers care for their fellow human beings; it would have shown that philosophy is more than an autistic concern with empty generalities, that it is a moral and political force that must be taken into account; and it would have taught the younger generation, better than any government-supported philosophy program, that devoting some time to its study is worthwhile.

# INDEX

Abstraction, viii, 5, 16, 153, 195, 260; and Parmenides, 66, 67, 201; tendency towards, 35–40, 53, 198, 227–28, 258, 262

Abundance, viii, denial of, x, 14–15, 16, 212; loss of ability to preserve, 54; of the world, 3, 241

Academies, art, 97, 230

Achilles, 19–20, 21, 23–24, 27, 39n.32, 127, 253; and actions as socially conditioned, 185–86; and dichotomy of honor and independent "real" nature, 35–38, 179–85; as example of ways of looking at traditions, 83–85, 86; and "existential" element, 71; and foreboding use of analogy, 124, 126; and "inside view," 33, 37, 113, 114–15, 145, 183; and Parmenides, 67

Additivity, 23, 30

Aeschylus, 46, 52n.28, 53, 207

*Against Method*, viii, ix, and Copernicus, 75–76

Agamemnon, 19, 23, 35, 179

Aggregates, 24–25, 27, 30, 38, 179, 184

Alberti, Leon Battista, 96–97, 104, 107, 113, 229

Alkaeus, 45

Alkmaion of Kroton, 48–49

Ambarzumjan, Viktor, 149, 191

Ambiguity, viii, xvii, 63, 196, 203, 261; of cultures, 78, 123, 240; as an essential companion of change, 39n.39, 59

Amenophis IV, 34, 108–9

Analogy, 124, 126

*Analyse der Empfindungen*, 209

Anaximander, 43, 60, 64, 66, 87, 132, 156

Anaximenes, 42, 156

*Ancient Medicine*, 190

Andaluza, Gloria, 33n.25

Animism, 8

Antal, Frederick, 34

*Antilogike* (word bashing), 72–73

Antisthenes, 235–36

Apocryphal Acts of Saint John, 109–10

Apollo, 207

Appiah, Kwame Anthony, 267, 268

Approaches to interpretation, 83–88

Archilochus, 43–44, 64

Argument, 15, 58–59, 71, 73–74, 79; effects on the premise of, 63, 74–75; effects on from developments outside, 71nn.15, 16, 75

Aristotelians: and close contact with experience and objects, 148, 191, 231; and multiple essences, 7; worldview of, 152

Aristotle, 99, 131, 137, 186, 217–22; and aggregates, 38; and change and subdivision, 71, 218; and common sense, 219–20, 266; and continuity, 66; as empiricist, 237; and the Good, 69, 189, 200–201, 217–18, 248–49; historical acceptance of, 220–21; and imitative view, 91; and "intermediate subjects," 190–91; and music, 29–30n.21; and Parmenides, 66–67, 69, 86, 200–201, 208, 248, 266; and potentiality and actuality, 170; and proof via description of worldview, 170; and reality as what is basic in life, 157, 208, 218, 248; and regularity, 7, 165, 192; and science as politics, 219–20, 222; and the vacuum,

Aristotle (*continued*)
221–22; and Xenophanes, 41, 43,
50–51
Aristotle's principle, 201–5
Aristoxenus, 29–30n.21
Arp, Halton, 149–50, 155, 191, 231
Art: as a product of nature, 223, 224,
225–26; and science, 96–97, 223–24,
230, 232, 240
Artaud, Antonin, 101
Artifact: created by science and art,
232–34; nature as, 234–41
Artists: controlled by "stage set," 104;
and questions of reality, 138; social
standing of, 97
Arts: ancient Greece and diversity of the
arts, 254; and reactions of contempo-
raries, 115–17; and science, 223–24; and
structure and projecting mechanisms
of the stage around them, 102, 117–22;
and styles, 101–2, 109
Astronomy, 132
Athenaeus, 46, 47
Athletes, Greek, 46–47
Atomism, 6, 68–69, 74, 157, 218
Auerbach, Eric, 25–26n.13, 178
Augustine, Saint, 126, 229

Babylonian astronomy, 132
Bacon, Roger, 166n.8
Baptisterium, 94–95, 99
Baxandall, Michael, 27–28
Beauty and Goethe, 225
Beginning and Parmenides, 87
Behaviorists, 120
Being: based on Aristotle's principle,
204–5; conservation of, 61, 66; ground
of, and Xenophanes' conception
of God, 54; independent of any
approach, 205; interaction with, xi,
xviii; and 'not being', 17, 65, 69–70,
156, 187–88, 198–99, 218; and Parmenides, 60–79, 148, 168–69, 187–88,
198–99, 218–19; proof of uniformity
of, 15; unity of, and Western intellectual life, 66; and Western science,
245–46
Bell, John S., 76, 171–72, 197, 205

Bible, 4–5n.3, 56, 168, 178, 196
Blocking of input, 4, 5, 13
Boccaccio, Giovanni, 117n.25
Bohm, David, 204
Bohr, Niels, viii, 71n.15, 76, 143, 144, 197;
complementarity, 172; and dialogue
with Einstein, 267; and political
interventions of scientists, 173; and
separation of the experimentalist, 203,
204
Boltzmann, Ludwig, 162
Born, Max, 231
Borrini-Feyerabend, Grazia, vii, xv
Boundary between collective outlook and
"the world," 27
Brahe, Tycho, 6n.6, 7, 165, 193, 236
Brouwer, Luitzen, 57
Brunelleschi, Filippo, 32n.23, 107; experi-
ment of, 94–100, 113; painting inter-
preted as a stage, 100–104, 111, 113–15,
120
Burbridge, C., 151
Burkhardt, J., 106n.19

Camera obscura, 100, 102
Cartesian method. *See* Descartes
Catastrophes, and creator god idea,
164–65, 192–93
Ceausescu, Nicolae, 250, 251, 262
CERN, 103, 144, 145, 216
Chambers, Robert, 149
Change, vii, 219, 253, 266; and
Democritus and Aristotle, 71, 266;
devaluation of in nineteenth-century
science, 156; and Empedocles, 190;
and existential element, 72; and
Parmenides, 50, 60, 66, 68–69, 156,
170, 187–88, 199, 208, 218
Chargaff, E., 142
*Chemistry, Quantum Mechanics, and
Reductionism*, 249
Chinese thinkers and diversification, 7,
165, 192, 237
Christ: body and blood of, 115; debates
about nature of, 168; divine nature of,
111n.16
Christian art, early, 93
Cleisthenes, 37, 183, 227

*Common Knowledge* (magazine), xv, xvi, 267, 268
Common sense, 10, 144, 208, 219, 253; and Aristotle, 69; and Arthur Fine, 207; Athenian, 252–53, 260; and Leucippus, 70; and Parmenides, 99
Concealed nature of world, 11
Concepts, 257, 271
Conclusion, and its effect on the argument, 74–75
*Conquest of Abundance*, vii–viii, ix, xv
Conservation of Being, 61, 66, 156, 169, 188, 199, 219, 248
Constantine V, 92
Constantius II, 106
Continuity, 66
Copernicus, Nicolaus, 72, 74–75
Corneille, Pierre, 55
Cornford, M., 42
Cosmology, 149, 151, 155, 231
Counterexamples and argumentation, 68, 76–78
Creativity, 226, 228–29, 230
Creator god idea and catastrophe, 164–65
Criteria of existence and scientific practice, 136–37
Crude ideas, 13–14, 260
Cultural: impoverishment and philosophy and science, 269–73; periods and lack of uniformity, 34–35
*Culture and Truth*, 33–34n.25
Cultures: closed, non-existence of, 78; and guaranteeing of well-being of members, 212–13; interaction of, 263, 266–68; nature of, 85; openness of connected with ambiguity, xi, 78; as potentially all cultures; xi, xvi, 33, 33–34n.25, 215–16, 240–241
Curtis, Heber, 149, 191, 231

Darwin, Charles, 72, 245
Delbrück, Max, 148, 247
*Della Pittura*, 96
Democratic, universe as, 242, 243
Democratization and transition to abstraction, 227–28
Democritus, vii, 99, 157, 170, 201–2n.7,
222; change and Parmenides, 68–69, 71, 170; and paradox of discernment of reality, 15n.15; and sensations and feelings, 246–47; and world as illusion, 11
Denis, Saint, cathedral of, 214
*De Revolutionibus*, 75–76
Descartes, René, 6, 84, 117–18, 140, 142, 202, 236; and laws of nature, 219, 237
Details, 14–15, 27. *See also* Particulars
Deuteronomy, 4–5n.3, 56
Development of transformation in Greek culture, 26–27, 31, 35–340
Dichotomy, 9–10, 13, 173–74; crudity of, xvii, 184; of Good and Evil, 5; and Homeric thought, 182; nature of, 167–68; real/apparent, 5, 13, 16, 17–18, 22, 36, 39, 178; subjective/objective, 143
*Die Ausdrücke für den Begriff des Wissens in der vorplatonischen Philosophie*, 253
Dieks, Dennis, 147
Diels, H., 45, 47
Diomedes, 19, 270
Dirac, Paul, 249
Divine judgment, 36–37
Dreams, 9, 31, 178
Drinking, 45–46
Duhem, Pierre, 220
Dyson, Freeman, 141

Earth: dynamics of motion of, 75–76; and its depth, 42–43
Egyptian art, lifelike versus stiff, 107–9
Ehrenhaft, Felix, 161–62
Einstein, Albert, 149, 156, 161, 162, 191–92; as a rational man, 172, 197; and Bohr, 267; and definition of realism, 171; as empiricist, 11n.16, 16, 62–63, 148, 199–200, 231, 266; and general relativity, 141, 219; and illusion of time, 16, 60, 188, 198, 266; and music, 30n.21; and special relativity, 245; and world as illusion, 11, 16, 169–70, 188
Einstein-Podolsky-Rosen correlations, 68, 76, 144, 197, 199–200

Eleatics, 69, 187, 189. *See also* Leucippus; Parmenides; Zeno
Eliade, Mircea, 51–52, 227
Emerton, Norma, 138n.13
Empedocles, 18, 190
Empiricism, 7, 193, 199, 237
English language, 123–24
Environment, degradation of, 8
Epistemologies, and science, 231–32
*Erkenntnis und Irrtum,* 228–29
Escher, M. C., 85
Ethics, 7–8, 242–51
Euler, Leonid, 150, 155
Euripides, 207, 259
European Economic Community and economic reform, 250–51, 261–62
*Euthydemus,* 58
Evans-Pritchard, E. E., 123–24
Existential feature of discussions about reality, 71–72, 202
Exner, Franz, 162
Experience, nonreliability of knowledge from, 61

Faces, 209–10
Facts, isolated, 47
Falsification, 68, 77–78
Fang Lizhi, 242–45, 249–51
Fanon, Frantz, 177
Faraday, Michael, 117–18
*Farewell to Reason,* xi, 30
Feyerabend, Paul, ix–xii
Feynman, Richard, 141n.23
Fierz, Markus, 164
Filters, xi, xii
Fine, Arthur, 84, 133n.5, 207
*Florentine Art,* 34
Formulae, 22
Frank, Joseph, 267, 268
Fränkel, Hermann, 45, 49
Freud, Sigmund, 162, 214
Fukuzawa, Yukichi, 160

Galileo, Galilei, 6, 9, 72, 76, 221, 237; and explanatory talk, 126–27; and sensations and feelings, 246–47
Geller, Margaret, 149–50, 155, 191
General relativity theory, 141

Genesis and realism, 168
Geometric period: and Homeric world, 27–35, 252; knowledge in, 47–51; language in, 26
Geometry and painting, 96–97
Gershenson, D. E., 186–87
Ghiberti, Lorenzo, 104
Gibran, Kahlil, xviii
Giotto, 116–17
Gnosticism, 11, 13, 168, 169, 211, 238, 247
God: development of conception of, 226–27; finger of in natural anomalies, 7, 164–65, 192–93, 236, 245; as ineffable, 196–97, 214, 233; of Judaism, Christianity, and Islam, 139–40, 211; names of, 195–96; and natural laws, 165, 193, 202, 236–37; proof of, 36; and Xenophanes, 51–55, 63, 134–35, 168, 227
Goddess and Parmenides, 87, 247
Gods, 134, 139; anthropomorphic, 30–31, 136; as illusion or real, 246–47. *See also* Greek Gods
Goethe, Johann Wolfgang von, 225–26, 229
Good and Aristotle, 69, 189, 200–201, 217–18, 248–49
Gothic cathedrals, 233–34
Greek: astronomy, 132; language, 21–27; morality, 259–60; philosophers, 185–86; philosophy and science, origin of, 13, 14; religion, 168–69; virtue, 258–60, 270
Greek Gods, 9, 52–53, 226–27; loss of individuality of, 226–27; and proof, 57–58, 133–39. *See also* Gods
Greenberg, D. A., 186–87
Greene, Graham, 3
Grinevald, J., x
Grosseteste, Robert, 229
Guericke, Otto von, 221
Guthrie, W. K. C., 42–43, 51, 227

*Harmonielehre,* 226
Heavens and Xenophanes, 49–51
Hecataeus, 43
Hector, 23

Hegel, Georg Wilhelm Friedrich, 214, 227
Heisenberg, Werner, 148, 175, 192
Heitsch, E., 41–42
Heraclitus, 31, 67–68, 263
Hermeneutic frontiers, 267
Hero, Homeric idea of, 44
Herodotus, 46, 52n.28, 133–34, 260
Hesiod, 46n.14, 52, 86, 133–34, 186
History, as producer of knowledge, 132–33, 136, 144, 145–46
*History of European Thought in the Nineteenth Century, A,* 193
Homer, 43–44, 46, 52, 84, 133–34, 252–53; dreams as objective, 31, 178; Gods of, 52n.28, 54, 57, 133–39, 226; quoted by Socrates, 67–68; thought and grand subdivisions of, 36, 182; and virtue, 259, 270–71; worldview of, 27–35, 36
Homeric Greek: Achilles modification of, 183; and development beyond, 26–27, 184–85; equal reality of, 21–27
Honor, 19–21, 35–36, 178–79, 181–82, 206
*Hopi Genesis,* 243
Hovelaque, 3
Hoyningen, Paul, 144n.28
Hubble, Edwin Powell, 149, 155, 231
Human genome project, 211, 239
Humans as aliens in a strange world, 167–69
Hume, David, 137
Husserl, Edmund, 254
Hydrodynamics, 150, 244

Idealism, 197
Ideas, 6, 17, 58, 59
Ideology, 120
*Iliad,* 19–27, 35–38, 54, 64, 178–85, 260, 270–71
Imitative view, 89, 91, 92, 97–98, 114, 115
*Imperial Eyes,* 12n.17
Incommensurability, 26–27, 31, 37, 255, 267
Infinity, 42–43

Information: necessary to appreciate artwork, 119–20; obtaining and justifying of, 253–54; as phenomena and realities, 58–59
"Inside view" and new realities, 33, 37, 113, 114–15, 145, 183
Instrumentalists, 143, 157, 166, 193, 202
Intellectualism and violence, 5–6
*Invention of America, The,* 145
Ionian cosmologists: and unity behind diversity, 187, 198. *See also* Anaximander; Anaximenes; Thales

Jacob, François, 3
Japanese and introduction of science, 160
Jason, 206–7
John (apostle), 109–10
Jokes and Xenophanes, 41–42
Jung, C. G., 174–75, 214

Kant, Immanuel, 16–17, 29–30n.21, 84, 229
Katz, Stanley N., 267, 268
Kepler, Johannes, 6n.6, 7, 91, 159, 174; and telluric soul, 165, 193, 236
*Killing Time,* x
Kirk, G. S., 42
Knowledge, 7–8, 17, 55, 61, 196; and community, 257; Divine, 47; inaccessibility to humans, 47–48; late geometric, 47–51; and perception, 67–68, 70, 73–74; as result of historical development, 131–32, 136, 144, 145–46; Socrates' theory of, 67–68, 254–55; in *Theaetetus,* 254–60
*Knowledge and Passion,* 268
Kokoschka, Oskar, 116
Koshland, Daniel, 211
Kranz, W., 45, 47
Krautheimer, Richard, 95, 96
Kuhn, Thomas, 33n.25, 144n.28, 267

Lactantius, 75
Language, xvii, 20, 21–27, 144, 181, 227; as bearer of worldviews, 27–28, 31–33, 78; connection between "inside" and

Language (*continued*)
  "outside" of, 33, 37, 113, 114–15, 145,
  183; and the field of experience,
  28–29; limits of, 20, 180; "misuse" of
  by Achilles, 179–82. *See also* Homeric
  Greek
Laplace, Pierre-Simon, 238
Lamarckism, 148
Laws, Dr., 116n.24
Laws: fundamental, lack of a consistent
  set of, 141–42; postulation of univer-
  sal and inexorable set of, 6–7, 201,
  237
Leibnitz, Gottfried Wilhelm von,
  166n.8, 202, 219, 237
Lenard, Philipp Eduard Anton von, 161
Leonardo da Vinci, 91, 98, 99, 111n.21
Lessing, Gotthold Ephraim, 55, 211,
  213
Leucippus, 17, 68–70, 157, 189
Levins, R., 142
Lewontin, R. C., 142
*L'homme*, 117–18
Lists, 21, 254, 257, 259, 260, 271
Logic, xvii–xviii, 15, 58, 187
Lorentz, Hendrik, 161
Lorenz, Konrad, 155, 245
Luria, A. R., 4
Luria, S. E., 148, 149, 191, 230–31
Lycomedes, 109–10
Lydians, 45–46, 47

Mach, Ernst, 17, 162, 209, 228–29
*Madonna delgi Occhi Grossi*, 89, 90, 93
*Madonna del Granduca*, 89, 90, 91, 93
Maestlin, Michael, 6n.6
Mandelbrot, B. B., 119n.27
Manetto di Jacopo Ammanatini, 99n.9,
  101
Manifest world as illusion, 11
Marcion, 211, 247
Margaritone d'Arezzo, 90–91, 92
Marxism and world as illusion, 11
Masks and stereotypes of expression, 28
Mass, 115
Mathematics, 47, 260n.7, 263–64; pure,
  and art, 93, 117; and *Theaetetus*,
  255–58

Matter, transformation of, 233–34
Maxwell, James Clerk, 150–51, 161, 191,
  193, 232n.12
Mayer, Robert, 219
*Mbisimo*, 123–24
Medawar, Peter, 250, 252
Medea, 206–7
Melissus and Xenophanes, 43
Memory, 21, 73, 74
*Meno*, 258, 259, 260, 262
Merz, Johann Theodore, 152–54, 155, 193,
  194
Metaphysical hypotheses, 244–45
Meter, 21
Milesians, 42–43
Mill, John Stuart, 125
Mind-body problem, 141
Modern art, 117
Modern world, 7, 8, 250–1, 176–7, 242,
  261–2
Modified separability assumption, 136,
  138–39
Molecular biology, 142
Money, 28, 258
Monod, Jacques, 5–6, 7–8, 9
Monteverdi, Luigi, 91
Morphology, 153, 155
Motion, 23, 27, 70, 189, 228
Murchison, Roderick, 245
Murray, Gilbert, 53, 227
Muses and Hesiod, 87, 186
Music: as a way of knowledge, 229; and
  creation of moods, 28; Kant's valua-
  tion of, 29n.21, 229; and nature, 226;
  and Plato, 194
Mystics and Parmenidean One, 71
Myths, 218, 220

Naïve realism, 113, 122–23, 219
*Nathan der Weise*, 211
National Endowment for the
  Humanities, 224
National Science Foundation, 224
Nature, 6–7, 224, 234–41
Near Eastern law, 56, 186n.7
Needham, J., 131
Newton, Isaac, 140, 148, 159, 192, 231,
  261n.10; and finger of God, 7, 165, 193,

236, 245; and gravitational law, 226, 238, 261n.10

Nietzsche, Friedrich Wilhelm, 31, 227

*1984*, 14–15

Nirvana, 249

Normative philosophers, 84, 85

Not-Being, 17, 65, 69–70, 156, 187–88, 198–99, 218

Numbers, 8n.7, 263–64

Objective: account of human affairs, 120; physical laws and human vision, 102; side of experiment, 100, 103, 146

Objectivism, 6, 8, 163–64, 241

Objectivity, vii, 7, 76–77, 143. *See also* Objective; Objectivism

Odysseus, 19, 178, 179, 270

O'Gorman, Edmondo, 145

One, Parmenidean, 70–71

*On Human Nature*, 163

*On Melissus, Xenophanes, and Gorgias*, 56, 135–36

Ontological discussion, necessarily precedes arguments, 77

Ontological pluralism, 215

Opinion and Xenophanes, 49, 51

Optical realism, and spirituality, 110–11

*Oresteia*, 207

Orphism, 247

Ostwald, Wilhelm, 219

Painting: as imitative, 89–91; philosophy and content of, 229–30; and scientific principles, 96–97

Paratactic feature, 22, 47

Parmenides, xviii, 60–79, 86–87, 99, 186, 267; and atomism, 157; and Being, 60–79, 148, 168–69, 187–88, 198–99, 218–19; and denial of change and difference, 50, 60, 68–69, 156, 170, 187–88, 199, 208, 218, 248, 265–66; and establishment of objective reality, 190; and the Goddess, 87, 247; and Leucippus, 68–69, 189; and life as illusion, 16, 270; and movement towards abstraction, 262; "path of truth" of, 64–67; and proof, 63, 64–67; and transformation of

Greek culture, 31; and Xenophanes, 41

Parry, A., 31, 34, 35, 37; and defined rules, 38–39; and language of Achilles, 20, 179–80, 181

Particulars, 250, 252. *See also* Details

Pater, Walter, 229

Patterns and cultural forms, 28

Pauli, Wolfgang, 164, 172–76, 203, 214; and C. G. Jung, 174–75; science and salvation, 174–76, 177

Perception and knowledge, 67–68, 70, 73–74, 189

Performance of modern science, 6, 7, 9

Perry, Matthew Calbraith, 160

Perspective: as a stereotype, 107; and development of explicit and hidden, 113–14; development of in Renaissance, 98–99

Petrarca, 117n.25

Philosophy: approach of, 84; and cultural impoverishment, 269–73; and generality, 265; and Parmenidean One, 70–71

Physics, 62, 70, 141; and psychology, combination of, 214; and real world, 169, 188, 199

Pirandello, Luigi, 116, 210

Planck, Max, 11, 161, 204n.8, 221; and immediate sense impressions, 199–200; and problem of physics, 62–63

Plato, 11, 109n.20, 118, 154; and *antilogike*, 71–72; and the arts, 229, 230n.7, 270; and battle between philosophy and poetry, 31; and imitative view, 91; and knowledge, 67, 131, 270; and *Meno*, 258, 259, 260; and Parmenides, 66–67; and proof stories, 58; and realistic worldview, 170; and the "subjective," 254; and *Theaetetus*, 254–60; and *Timaeus*, 243; and unchanging real world, 194–95; and universals as tyrants, 252, 264; and use of the dialogue, 186

Platonists, 189, 201, 217, 248; and divine nature of numbers, 235–36, 237; worldview of, 152

Plutarch, 44, 46n.14
Podolsky, B., 68, 76
Poincaré, Jules Henri, 148, 192, 231, 245
Politics, science as, 219–20, 222
Pollock, Jackson, 230
Popkin, R., 165–66n.8
Popper, K., 52
Portraiture, 104, 106
Positivism, 197
*Power and the Glory, The,* 3
Pragmatists, 84
Prandtl, L., 150, 155, 191
Pratt, Mary Louise, 22n.17
Pre-Socratics, 56, 131
Primas, Hans, 249
*Principia,* 238
Progressivism, 93
Projections, xvii, 100, 101–4, 117–22, 143
*Prolegomena,* 16–17
Proof: as culturally dependent, 55, 57; on the eternity and unity of God, 56n.33; and Parmenides, 63, 64–67; and Xenophanes, 55–59
*Prophet, The,* xviii
Protagoras, 67–68, 73, 215n.9, 254, 262, 271
*Proverbs in Prose,* 225
Pryce, Maurice, 162
Przibram, Karl, 161
Pseudo-Dionysius Areopagita, 195–96, 214
Psychology, 174–75, 214
*Public Understanding of Science,* 157–58
Pythagoras and Pythagoreans, 70, 118, 186, 247

Quantum mechanics, 74, 128, 131, 141, 150; and Einstein-Podolsky-Rosen paradox, 68, 76, 144, 197, 199–200; its worldview and world problems, 176–77; and Parmenides, 203, 249; and properties depending on approach chosen, 213; and Socrates, 68; and subjective/objective dichotomy, 143; and symbolic terms, 175, 213

Quantum theory, vii, 158, 240; historicity of its assumptions, 142; and singleness of the world, 71n.13
Quarks, 136

Rainbow, 10
Raphael of Urbino, 89, 98n.8, 99, 230
Rationalism and rationality, xvii, 16, 184, 212, 213, 241
Raven, J. E., 42
Realism, 118–23, 127–28, 178–96, 215; as a worldview, 169–72; historical comments on, 197–205; and religious fervor, 247. *See also* Reality-appearance distinction
Reality, vii, viii, xvii, 5, 9–10, 58, 118, 211; different scientific approaches to, 191–95; "existential" features of discussions about, 71; as free of change, 66, 69; hidden behind appearance, 167, 200; ideas declared to be, 210–11; as ineffable, 211–14, 233; linked to senses according to science, 207–8; as material processes, 204, 207; as open and changeable, xviii, 79; and Parmenides, 66, 190; and perspective, 99–100; questions of, 209–11; and scientific viewpoint, 27, 207–8; and search for, 5, 10; as stage set, 104, 111, 113; three notions of in 5th to 6th century B.C. Greece, 190; as what is basic in life (Aristotle's principle), 157, 201–5, 248
Reality-appearance distinction, 16–18, 22, 36, 39, 178, 202; history of, 5, 198; and Parmenides, 61, 64
Reductionism, 140, 141, 154
Refutation, 78
Regularity, 7, 21, 165, 192, 193
Reinhardt, K., 42, 43, 49
Relativism, xi, 115–23, 127–28, 143, 202–3, 262; and ontological pluralism, 215
Religion, 4, 9, 10, 115–16, 219; and different levels of being, 218; and effects of various styles, 29; movements of, 70–71, 201; objects of,

115–16; and reality view fixed by fiat, 212; and science, 158, 159, 162, 164–66; worldview of, 246–47. *See also* God; Gods; Theology
Results, 120
Riegl, Alois, 93
Rise of rationalism, xvii, 16, 26, 44, 67, 184, 212, 213, 227
Roman Church, 9
Rosaldo, Michelle Z., 268
Rosaldo, Renato, 33–34n.25
Rosen, N., 68, 76
Rosenfeld, Leon, 162
Rules and changing worldviews, 38–39
Rumi, Maulana Jajal al-Din, xii–xiii

Santa Maria del Fiore, 94–95
Santillana, Giorgio de, 98–99n.9
Schoenberg, Arnold, 226
Schopenhauer, Arthur, 31
Schrödinger, Erwin, 71n.15, 161, 171, 219, 245; and Aristotle's principle, 201n.7
Schrödinger's cat, 128, 249
Schwitters, Kurt, 240
Science: and art, 96–97, 223–24, 230, 232, 240; and artifacts, 237–41; and common sense, 219; and concealing of origin of knowledge, 144; and different approaches to reality, 191–95, 202–3, 230–32, 239; facts depend on historical-political circumstances, 132; and God, 139–40, 142; and imitative view, 91; and inexorable laws, 6–7, 201, 237–41; and its ethic, 7–8; limits of its knowledge, 139–43, 145–46; and mind-body problem, 141; and naturalism, 163; no uniform enterprise of, 159, 232; and the path of least resistance, 142; and popularity, 157; and positing of "real world," 265–66; and practical advantage, 158; and projection of objective existence of the world, 115; and rationality, 212; and reality linked to the senses, 207–8; and reductionism, 140, 141; and results and ideology, 120; rise of, 26, 174; and status, 157–58, 159;

and success, 6, 203, 212, 213, 239; and supposed illusory worldviews, 246–47, 248; universal application of, 242, 244, 250, 264; worldviews of, 147, 152–55, 158, 159, 247. *See also* Scientific method
*Science in a Free Society*, xi
Scientific experiments: Brunelleschi's, 94–95; and the stage, 102–3
Scientific measurement, 237–38
Scientific method, 136–37, 144, 146, 148–52, 158
Scotus, Duns, 166, 193, 202, 236
Separability assumption, 133, 136, 139, 144
Sequences, and Greek language, 23–24
*Shaky Game, The*, 133n.5
Shapley, Harlow, 149, 231
Sherashevsky, 4
Simplification, 5, 12
Smith, Cyril Stanley, 138n.13
Snell, Bruno, 25n.12, 26n.14, 253, 258
Social refutation of a logical point, 67
Socrates, 38, 76, 254–56, 257, 258, 260, 271; and theory of knowledge, 67–68, 74–75, 76
Solon, 37, 227
Sophists, 68, 88, 186, 259
Sophocles, 52n.28, 182n.4, 259
Soul: Azande definition of, 123–24; lack of term in Homeric language, 25; and optical image, 110–11
Sparta, 44
Spinoza, Baruch, 236
Spirituality and optical realism, 110–11
"Spontaneous tolerance," xi
Stage: Brunelleschi's painting interpreted as, 100–104, 111, 113–15; as reality, 104, 113–14, 120, 122
Stanislavsky, Konstantin, 91
Stark, Johannes, 161
Stereotypes, xvii, 104, 106, 123
Stesichoros, 46n.14
Stevin, Simon, 228
Stoics, 201
Structure, 39n.32, 78–79, 181, 184
Subjective side of experiment, 100, 103, 156

Subjectivity and Achilles, 38, 184

Substance, as principle explaining diversity, 60–61

Suger of Saint-Denis, Abbot, 29n.20, 233–34

Superstring theoreticians, 141n.23

Surroundings as reality, 212

Systematic analysis, vii, 154, 224

Terpstra, Bert, ix–x

Thales, 42, 49–50; and water substance, 60, 156, 187, 198

*Theaetetus*, 67–68, 73, 75, 254–57, 260

Theodorus, 256, 257

Theology and natural laws, 165, 193, 202, 236–37

Theory, 151–52, 252, 261

*Theory of Colors*, 225

Thirring, Hans, 161, 162

Thomas, Saint, 166, 193, 202

Thomson, J. J., 161

Thutmosis, 108

*Timaeus*, 243

Time-independent laws, 140–41

Timon of Phleios and Xenophanes, 54, 135

Toricelli, Evangelista, 221

Tradition, 58, 61, 261; and learning of virtue, 270–71; nonreliability of knowledge from, 61; stage and means of projection as, 120, 122–23; three ways of looking at, 83–88. *See also* Cultures

Tragedy, Greek, 55, 259

Travelers and language, 12, 32

Truesdell, Clifford, 155

Truth, love of, vii, 158

Tyranny of the particular, 250, 252

Uhlemann, B., x

Understanding as a process that transforms the subject, 12, 32

Uniform world, xvii, 16, 26, 44, 67, 184, 212, 213, 227

Unity: of Being, 66, 72–73; of God, 56n.33, 63, 187, 198; of scientific worldview, 141–42, 154, 167

Universals, 252, 262–64; in Homeric times connected particulars, 252; as mediators, 261, 263–64; as tyrants, 252, 261, 262–64

Universe, as example for humans, 243

Upanishads, 11

Ur-Nammu, 56

Vasari, Filippo, 89–90, 93, 104, 107, 108

Vienna Circle, 162

Violence and intellectualism, 5–6

Viollet-le-Duc, Eugène Emmanuel, 232–33

Virtue, 258–60, 270–71

Voegelin, Erich, 4–5n.3

Void, existence of, 17

von Fritz, Kurt: and cultural changes in Greece, 258; and Xenophanes' construction of the argument, 58–59; and Xenophanes' idea of God, 52

von Neumann, John, 249

Waking reality, 9

W and Z particles, 71, 103

*Wärmelehre*, 17

Webern, Anton von, 225–26

Weinberg, Steven, 164

Western civilization and economic transformation of old cultures, 250–51, 261–62

Weyl, Hermann, 66, 156, 266

Whitehead, Alfred North, 84

Wholeness: and lack of in Homer, 23–25, 30; and language, 27

Whorf, Benjamin Lee, 27

William of Ockham, 166, 193, 202, 236

Williams, Bernard, 25n.12

Wilson, E. O., 158–59, 163

*Witchcraft, Oracles, and Magic among the Azande*, 123–24

Wittgenstein, Ludwig, 84

Word bashing (*antilogike*), 72–73

Worldviews: and actions and perceptions of artists, 93; alternatives to science, 159; conflicting of Jason and Medea, 207; discussion of precedes argument, 77; and Homer, 27–35, 36; and lan-

guage, 31–33; other than science, 246–47, 248; Pauli's combination of science and salvation, 174–76, 177; and physics, 162; power of, 164–69; realism as, 169–72; religious, 164–66, 192–93; scientific, 147, 152–55, 158, 247
Writers and valuation of language, 29–30n.21

Xenophanes, 31, 41–59, 64, 86; and God, 52–55, 63, 134–35, 168, 227; and proof, 55–59; and social criticism, 45–47

Zeno, 66, 69, 189, 221
Zeus, 36, 52, 141, 182, 207